CLARENCE LAPHAM gained an MSc at the University of Wales. Following a few years in industry he taught chemistry and mathematics at Croesyceiliog and Fairwater Schools, Gwent. He later became head of the Upper School of Newport Bettws Comprehensive and then headteacher of Oakdale Comprehensive School, Blackwood, Gwent. Currently headteacher of Caerleon Comprehensive School, he was for many years an evening lecturer at Gwent College of Higher Education. Chief Examiner in general science and chemistry for two CSE boards, he also examines at GCE O-level.

key facts

Multiple Choice

BIOLOGY, J. M. Kelly, BSc

CHEMISTRY, C. W. Lapham, MSc

ECONOMICS, J. E. Waszek, BSc(Econ)

ENGLISH LANGUAGE, R. L. Wilson, MA

FRENCH, I. Bryden, BA

GEOGRAPHY, N. E. Law, BA

GEOGRAPHY, BRITISH ISLES, D. Bryant, BSc and R. Knowles, MA

HISTORY, SOCIAL & ECONOMIC, S. E. Haworth, BA and M. C. James, BA

HUMAN BIOLOGY, S. R. Cantle, BSc MMedSci

MODERN MATHEMATICS, A. J. Sly, BA

PHYSICS, B. P. Brindle, BSc

GCE O-Level and CSE Chemistry

C. W. Lapham, MSc

key facts

Published by Charles Letts and Co Ltd
London, Edinburgh and New York

Published 1983 by Charles Letts & Co Ltd
Diary House, Borough Road, London SE1 1DW

1st edition 1st impression

Made and printed by Charles Letts (Scotland) Ltd
ISBN 0 85097 567 0

Contents

Introduction

This book has been written as a companion to the very successful Key Facts O and CSE Chemistry Passbook.

Providing a very concise text, the Passbook gives information relating to the topics contained in current chemistry examination syllabuses, with each chapter of this book corresponding to a chapter in the Passbook.

As the title of this book implies, it has been written for students to test themselves using fixed response questions of the multiple choice variety. However, some chapters end with questions of the structured kind involving some free responses. These are included so that students can have the opportunity of testing themselves on a series of related facts in a single question as an appropriate end to their work on the particular chapter in question.

In short the Passbook provides the text, while Multiple Choice O-level and CSE Chemistry provides a very easy and convenient means whereby a candidate can check whether or not he/she has learned and understood the work. This book could be used independently of the Passbook providing as stated previously the means of testing work learned at school, from notes or any other source of relevant syllabus material.

Question types used in examinations

These can be split conveniently into two broad categories (a) fixed response questions and (b) free response questions.

(a) Fixed response

These are questions which have a set or fixed response to them, and are often described as multiple choice and objective type questions. There are several kinds of fixed response type questions, the more important being summarised and exemplified below.

1. **Simple multiple choice questions**. Here a question is asked and four or five possible answers are given, only one of which is correct.

e.g. An insoluble salt is prepared by the method of

A alkali + acid   B basic oxide + acid
C carbonate + acid   D metal + acid
E double decomposition

The correct answer is (E) because all the other answers (called distractors) are methods used for the preparation of soluble salts. This particular question involves the complete and total recall of facts, as opposed to other questions set which test knowledge, application and understanding.

2. **Classification questions**. These are what could be described as 'compound multiple choice' questions. Each question consists of a list of headings followed by a number of questions or completions. To answer, candidates must match up the letter of the heading with the most appropriate answer to the question or the most appropriate completion as the case might be. As in the case with examinations the best method of description is via an example.

e.g.

A nitric acid (HNO_3)   B hydrochloric acid (HCl)
C sulphuric acid (H_2SO_4)   D calcium hydroxide ($Ca(OH)_2$)
E potassium hydroxide (KOH)

Which one
1 produces a white precipitate with barium chloride solution?
2 reacts with magnesium to produce magnesium sulphate and hydrogen?
3 gives a white precipitate with carbon dioxide?
4 gives a white precipitate with aluminium sulphate which dissolves in excess?
5 gives a white smoke with ammonia gas?

The answers are 1C, 2C, 3E, 4D, 5B. As can be seen an answer can be used once, more than once or not at all.

As a matter of information and interest this type of question can prove to be versatile in that several aspects of the syllabus can be tested using the same question. Again let's look at the example. Further questions could be asked thus:

1 Which compound contains the greatest number of atoms?
2 Which acid could not be prepared by the action of another acid on one of its salts?
3 Which compound could be used to dry carbon dioxide gas?
4 Which compound could be used to absorb carbon dioxide?

Answers 1C, 2C, 3C, 4D.

In practice more could be asked regarding the other compounds in the list, but this particular example again serves to illustrate the fact that an answer may be used once, more than once or not at all.

3. **Multiple completion questions**. Here the question is asked and this is followed by a number of answers of which one, two or three can be correct (hence the name multiple completion). Using a code specified by the examination board the student then answers. Yet again an example is the best method of explanation.

e.g. Which of the following gases are elements?
1 nitrogen (N_2)
2 hydrogen (H_2)
3 sulphur dioxide (SO_2)

One of the codes most commonly used in order to answer is
A if 1, 2 and 3 are all correct
B if 1 and 2 are correct
C if 2 and 3 are correct
D if only 1 is correct
E if only 3 is correct

In this example both nitrogen (N_2) and hydrogen (H_2) are elements, hence 1 and 2 are correct so the answer is B.

It is only fair to stress at this point that there are variations to this type of question, e.g. some examination boards give four answers and use the coding
A if 1, 2 and 3 are all correct
B if 1 and 2 are correct
C if 2 and 4 are correct
D if only 4 is correct
E if some other combination is correct

4. **Assertion–reason questions**. At times, and more often at CSE level, questions are asked involving 50/50 type answers e.g. true/false or yes/no type questions. Obviously by the law of averages marks can be obtained easily by sheer guesswork and who is to blame a candidate for guessing, indeed it is virtually impossible for an examiner to prove guesswork on the part of a candidate. Assertion–reason questions are designed to minimise the possibility of guesswork by giving two statements, the candidate having to decide whether *each* is true or false according to the following code:

If both statements are true, and the second statement is a true explanation of the first then the answer is A.

If both statements are true, but the second statement is not a true explanation of the first then the answer is B.

If the first statement is true and the second statement is false then the answer is C.

If the first statement is false and the second statement is true then the answer is D.

If both the statements are false then the answer will be E.

e.g. When zinc is added to a solution of copper(II) sulphate copper is formed **because** copper(II) sulphate is a hydrated salt.

The answer is this case is B: zinc does displace copper from a solution of copper(II) sulphate, and copper(II) sulphate is a hydrated salt **but** the second statement is not an explanation of the first.

All of the types of fixed response questions described in this introduction are used by examination boards. Below is a summary of the extent of the usage. It should be borne in mind that in England and Wales there are 12 setting CSE examinations, 7 setting O-level examinations with the Welsh Board setting both CSE and O-level exams. Scotland and Northern Ireland also have their own examination boards.

Types of questions	CSE Boards	GCE Boards
Multiple Choice	9	5
Classification	8	5
Multiple Completion	2	4
Assertion-reason	1	2

These figures do not include the Scottish and Northern Ireland Boards.

(b) **Free response questions**

These can best be described as questions requiring the candidates to write an answer 'on their own' rather than make a selection from a list. Answers can vary from those requiring one word, e.g. completion of sentences, to writing an essay.

Three main types of questions would seem to cover the range of free response questions.

1. **Short answer questions** in which the answers are one word or just phrases, e.g. Name a halogen. Answer: chlorine or bromine or iodine etc.

2. **Structured questions** involving an initial statement or diagram or both and being followed by a series of questions that are related to the initial statement and which get progressively more difficult.

e.g. When the gas sulphur dioxide is bubbled into the water the solution formed is acidic and smells of the gas. When sodium hydroxide solution is added to the acid solution the smell of sulphur dioxide disappears.

(a) (i) Give a simple test to prove that the solution is acidic
 (ii) Write the formula for sulphur dioxide
 (iii) Name the acid formed in the reaction
 (iv) What kind of substance is sodium hydroxide solution?

(b) Explain why the smell of sulphur dioxide disappeared after adding a sodium hydroxide solution.

(c) How could sulphur dioxide be obtained from the solution formed in (b) etc.

Answers (a)(i) blue litmus or indicator paper turns red in acid solutions (ii) SO_2 (iii) sulphurous acid (iv) an alkali.
(b) the sodium hydroxide has neutralised the acid.
(c) Add a stronger acid. This would displace sulphur dioxide from the solution, etc. The question could be further extended.

3. **Essay questions** in which candidates would be asked to show an 'in depth' knowledge of a major topic e.g. the Periodic Table or the activity series etc.

How to use this book

1. Learn the work thoroughly from your school notes, text book etc. This can best be done by writing or using summary notes.

2. Work through the syllabus logically, topic by topic. Once you think you know and understand a particular section of the work, attempt the appropriate set of questions in this book, working your way through the questions sequentially.

3. Now check the answers. A score of 75%+ is excellent, while one of less than 45% would give cause for concern. Always go back and check on anything you have wrong.

4. Do not gallop through the objective items, give them sufficient thought. Multiple choice items look deceptively easy. Remember they are designed for candidates to give sufficient thought to answering the question. The writing takes very much less time than the thought process used to arrive at that answer. Keep an eye on the **time factor**.

Acknowledgements

I would like to express my thanks and admiration for the work of Mrs Pat Bosanko in typing and advising on the original manuscript. For the help, support, tolerance and understanding my thanks to my wife Brenda, my daughter Karen, and my sons Jeffrey and Lee. Finally to my colleagues at Caerleon Comprehensive School for their advice and encouragement.

Chapter 1
The States of Matter

Simple multiple choice

1.1 A common substance that sublimes is
A ammonium carbonate
B sulphur
C iodine
D bromine
E chlorine

1.2 When a bicycle tyre is being pumped up at constant temperature, the pressure increases because
A the particles of air become bigger
B more air particles are hitting the tyre
C less air particles are hitting the tyre
D the air particles possess a lot of kinetic energy
E the air particles are moving faster

1.3 The random motion of a particle inside a liquid commonly called Brownian Motion or Movement is due to
A bombardment of the particle by liquid particles
B convection currents
C the pressure inside the liquid
D an increase in temperature
E the size of the liquid particles

1.4 Which of the following is used when calculating the volume of a gas at constant pressure with increase of temperature?
A Boyle's Law
B Charles' Law
C Ohm's Law
D Gay Lussac's Law
E Avogadro's Law

1.5 When a metal conducts, heat is transferred by the
A particles in the metal
B electrons and particles in the metal
C protons and neutrons in the metal
D increase of potential energy of the particles created by heating
E decrease in the kinetic energy of the particles

1.6 The volume of a gas at 20°C and 750 mm of mercury is $250\,cm^3$. The expression for the volume of the gas at stp (standard temperature (273 K) and pressure (760 mm of mercury)) is

A $250 \times \frac{20}{273} \times \frac{750}{760}$

B $250 \times \frac{273}{20} \times \frac{760}{750}$

C $250 \times \frac{293}{273} \times \frac{750}{760}$

D $250 \times \frac{273}{293} \times \frac{750}{760}$

E $250 \times \frac{273}{293} \times \frac{760}{750}$

1.7 $500\,cm^3$ of gas took 50 seconds to diffuse through a porous pot. The same volume of hydrogen (molar mass = 2) took 10 seconds to diffuse under the same conditions. The molar mass of the gas is

A 10 B 50 C 100 D 250 E 500

1.8 The gas which diffuses fastest is

A hydrogen
B helium
C nitrogen
D carbon dioxide
E oxygen

Ar(H) = 1, Ar(He) = 4, Ar(N) = 14, Ar(C) = 12, Ar(O) = 16
or
H = 1, He = 4, N = 14, C = 12, O = 16

1.9 In which of the following do the particles have the fastest random motion?

A a solid at room temperature
B a liquid at room temperature
C a gas at room temperature
D a gas at 100°C
E a liquid at 100°C

1.10 If the pressure is increased on a liquid it

A changes into a solid
B decreases its boiling point
C increases its boiling point
D causes the particles to come closer together
E decreases the volume of the liquid

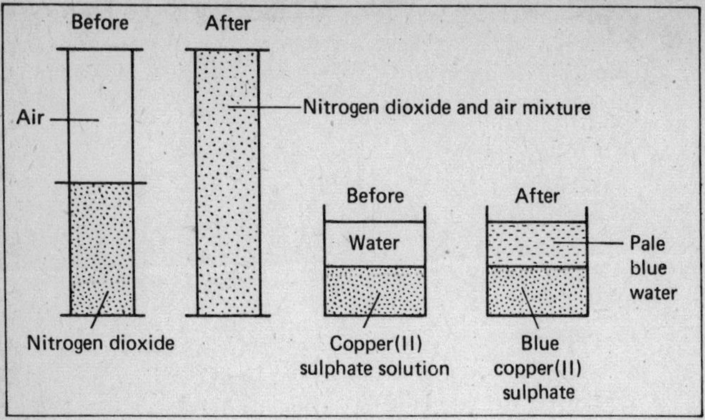

Figure 1

1.11 In the experiments illustrated in the diagrams (i) the gases have mixed completely (ii) the liquids have mixed slightly. This is because

A the particles in liquids are heavier than those in gases
B only gas particles move
C liquids cannot move in air
D nitrogen dioxide is less dense than copper(II) sulphate
E gas particles move faster than the particles in a liquid

1.12 Some students carried out an experiment to find the freezing point of a liquid (freezing point 7°C). They plotted the results on a graph. The correct graph is

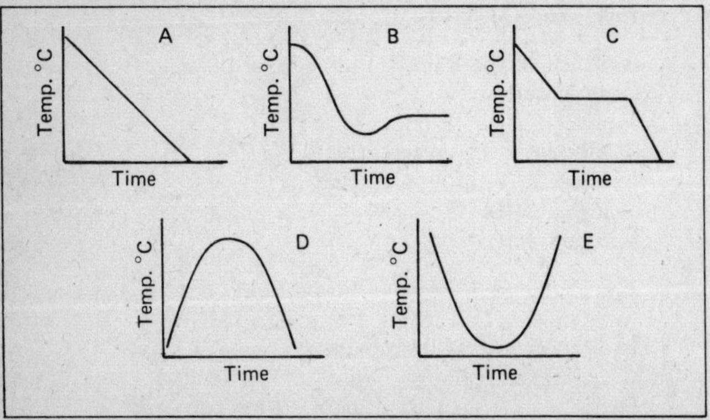

Figure 2

1.13 A drop of blue ink was dropped into a beaker of water. In a few hours all the water was coloured light blue. This demonstrates the process of

A osmosis
B dispersion
C convection
D reduction
E diffusion

Classification

A Melting
B Condensing
C Evaporating
D Subliming
E Diffusing

Which of the above are the best answers for each of the following (you may use an answer once, more than once, or not at all)?

1.14 Dry ice (solid CO_2) changing directly into a gas

1.15 Perfume being detected several metres away from the wearer

1.16 Perfume 'disappearing' when sprayed on to the body

1.17 Ice being changed into water at room temperature

A Solids
B Liquids
C Gases

Which of the above states of matter fit the following descriptions in the table?

	Volume	Density
1.18	Definite; takes the shape of the container	Medium
1.19	Definite volumes and shapes	High
1.20	Indefinite; takes the shape of the container	Low

Multiple completion

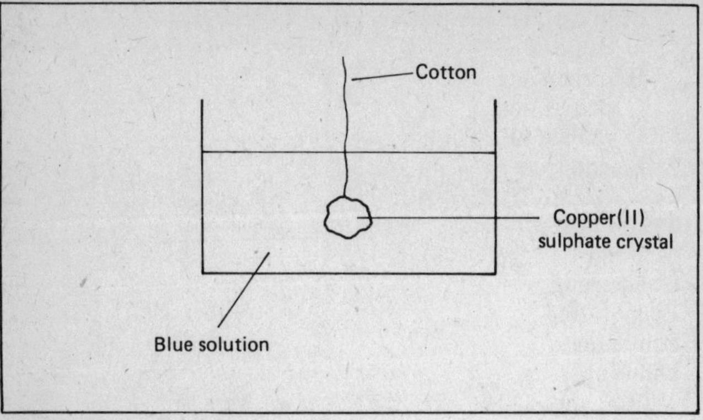

Figure 3

1.21 In the experiment above the solution went blue because of the process of
1 diffusion
2 melting
3 dissolving

1.22 The particles in a gas
1 are very far apart
2 are moving fast in random movement
3 have a high density

Assertion–reason

1.23 Heating a liquid increases the ratc of evaporation **because** The heat energy enables more and more of the particles to leave the surface of the liquid.

1.24 In the solid state the particles form irregular patterns **because** Particles in solids can move very fast and are confined to vibrations within their lattice structure.

1.25 When a liquid evaporates the most energetic particles leave the surface **because** The liquid is cooled by the process of evaporation.

1.26 When a liquid is cooled to below its freezing point it solidifies **because** When a liquid cools it loses kinetic energy.

Structured question

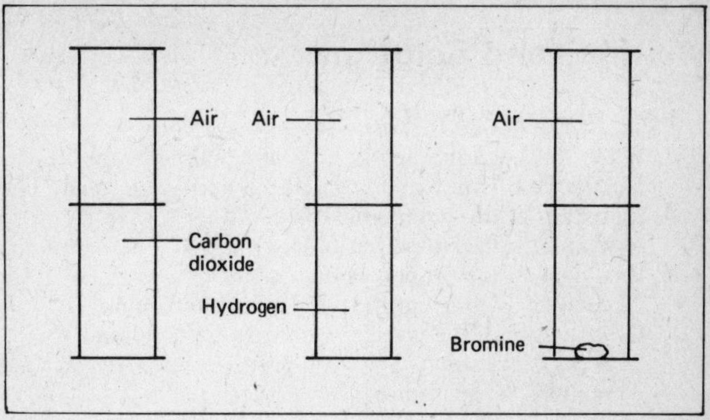

Figure 4

1.27 In the experiments above, the gases were allowed to come into contact with each other for about half an hour.

(a) (i) What would be the effect in experiment A of adding limewater to *both* gas jars?

(ii) What would be the effect in experiment B of testing each gas jar with a lighted taper? Give reasons for your answers.

(b) In experiment C the liquid bromine quickly evaporates and fills both gas jars.
What conclusions can you make about bromine from this observation?

(c) What general chemical principle is being demonstrated in these experiments?

(d) The molar mass of CO_2 is 44, H_2 is 2. What conclusion do you draw about the principle from these figures?

Chapter 2
Mixtures and Solutions

Simple multiple choice

2.1 When the two elements iron and sulphur are heated together the compound iron(II) sulphide is formed. The properties of this compound are
A similar to the properties of iron
B similar to the properties of sulphur
C completely different to those of iron and sulphur
D a mixture of the properties of iron and sulphur
E dependent upon the composition of the original mixture of the elements

2.2 In the following list the single pure compound is
A solder
B air
C tap water
D steam
E sea water

2.3 The main difference between a solution and a suspension is the
A solvents in which the solids are dissolved
B particle size of each solute
C particle size of each solvent
D 'thickness' of each solution
E temperature at which each was made

2.4 A chemical compound
A is two or more compounds chemically combined together
B is two or more elements mixed together
C is two or more elements chemically combined together
D contains only one kind of atom
E is always a product of a chemical reaction

2.5 When crude oil is fractionally distilled the first fraction is a mixture of various gases. The second fraction would
A burn very easily
B be a thick oily liquid
C be a black solid

D have a high density
E be a mixture of liquids and solids

2.6 A certain liquid was boiled and the vapour produced was condensed. The complete process taking place can best be described as
A polymerisation
B evaporation
C sublimation
D condensation
E distillation

2.7 In an aqueous solution of common salt the water is known as the
A solution
B solvent
C solute
D monomer
E hydrate

2.8 A saturated solution of a certain salt was made at 85°C. At this temperature a small amount of salt remained at the bottom of the solution. When this solution was cooled to room temperature (20°C)
A the solution cleared and no solid remained at the bottom
B the solution would no longer be saturated
C there would be a definite change in the colour of the solution
D there would be more salt at the bottom of the solution
E the whole solution would become solid

2.9 In an experiment to find the solubility of potassium nitrate per 100 g of water, at 25°C the following results were obtained

Mass of the evaporating dish	= 30 g
Mass of evaporating dish + saturated solution	= 37 g
Mass of evaporating dish + potassium nitrate residue	= 32 g

The solubility of potassium nitrate per 100 gm of water at 25°C is
A 2 g   B 5 g   C 20 g   D 40 g   E 80 g

2.10 The solubility/100 g of water of calcium hydroxide decreases with temperature. The graph which represents this fact is

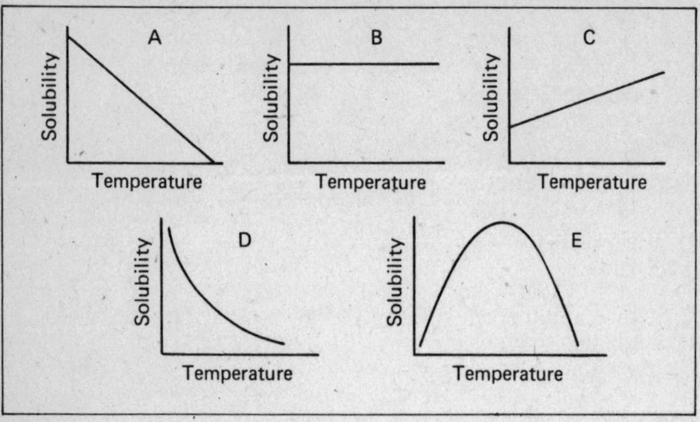

Figure 5

2.11 A solid element was heated strongly to form an oxide which looked very similar to the element itself. To prove that a compound is formed you should
A test it with indicator paper
B burn the oxide in oxygen
C examine the solid carefully under a microscope
D re-heat the oxide
E weigh the solid before and after the experiment

2.12 Some powdered glass and salt were mixed together. To obtain **dry crystals** of salt from the mixture the order of the process is
A mix with water, stir, evaporate part of the solution crystallise out, dry the crystals
B mix with water, stir, filter, evaporate part of the solution, crystallise out, dry the crystals
C stir, evaporate part of the solution, filter, crystallise out, dry the crystals
D mix with water, stir, centrifuge, crystallise out, dry the crystals
E mix with water, stir, filter, evaporate part of the solution, crystallise out

2.13 In processes used to separate substances use is always made of
A a difference in melting points and boiling points

B a chemical reaction
C a difference in a property of each substance
D a physical reaction
E the density of each substance

2.14 Water is a compound, proof of this fact is that water
A has a pH of 7
B boils at 100°C
C is formed when hydrogen burns in oxygen
D freezes at 0°C to form the solid ice
E reacts violently with sodium and potassium and forms two very strong alkalis

Classification

The diagram below refers to questions 2.15–2.19

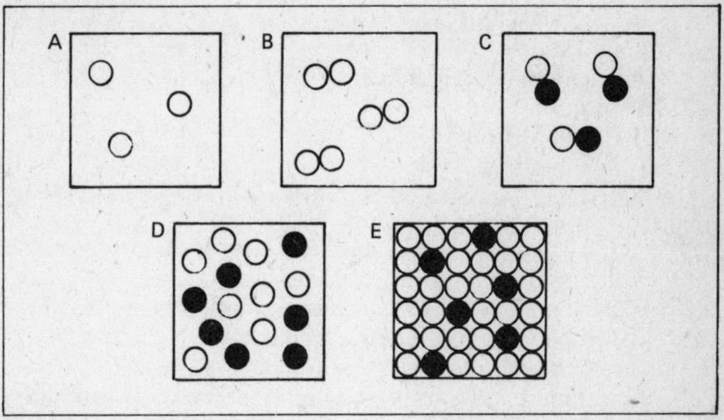

Figure 6

The circles (○) and dots (●) represent atoms of different elements. In answering you may use a box once, more than once or not at all.

Which box represents
2.15 an alloy?
2.16 a mixture of liquids?
2.17 a monatomic gas?
2.18 a diatomic gas?
2.19 a gaseous compound?

2.20 Substances can be solids, liquids or gases at room temperature, they can also be elements, compounds or mixtures.

Complete the following question by completing each part of the table

Substance	Solid (A) or liquid (B) or gas (C)	Element (A) or compound (B) or mixture (C)
1 Bronze		
2 Air		
3 Methane		
4 Copper		
5 Sea Water		

A Crystallisation
B Filtration
C Fractional distillation
D Solution
E Chromatography
F Sublimation

Select the correct process for each of the following (You can use an answer once, more than once or not at all)

2.21 obtaining the dye from a Smartie sweet
2.22 producing a sample of salt from a solution of sea water
2.23 separating the constituents of crude oil
2.24 removing solid material from reservoir water
2.25 separating a mixture of sodium chloride and ammonium chloride

Multiple completion

2.26 If pigments from green leaves are dissolved in ether and tested by using column chromatography the top of the column turns yellow, while a green band is evident at the base of the column. This experiment shows that the pigment
1 consists of two colours
2 consists of just one colour
3 is made up of yellow, green and blue

2.27 A solution is made up of
1 a solute and a solvent
2 a liquid and a solid that is soluble in the liquid
3 a liquid and a solid in fixed proportions

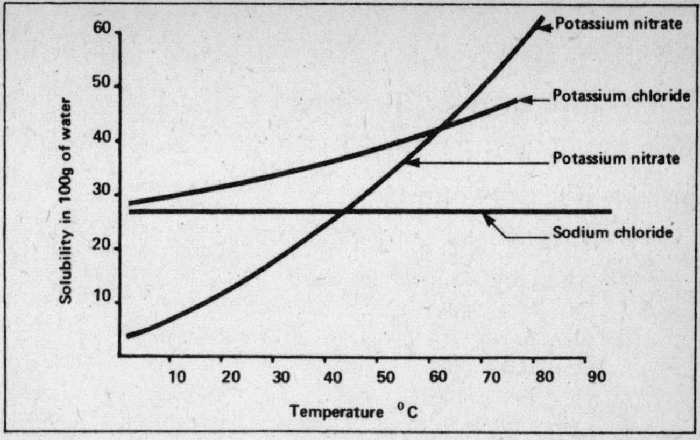

Figure 7

Questions 2.28 and 2.29 refer to the curves above.

2.28 The graph for sodium chloride shows that
1 the solubility of sodium chloride is fairly constant as the temperature increases
2 to prepare a saturated solution of sodium chloride heating would not help very much
3 30 g of sodium chloride dissolves in 100 g of water at 10°C

2.29 The solubilities of sodium chloride and potassium nitrate are
1 the same at 40°C
2 the same above 40°C
3 different about 40°C

Assertion–reason

2.30 Fractional distillation is a convenient process used to separate mixtures of liquids **because** In this process use is made of differences in chemical properties

2.31 Fast evaporation can be used to obtain sodium chloride from a solution of sodium chloride **because** Sodium chloride is unaffected by heat

2.32 Crystallisation from a solution is used to produce very small crystals **because** In the process of crystallisation the crystals have time to grow from the solution

Chapter 3
Formulae and Equations

Simple multiple choice

3.1 $x\,SO_2(g) + O_2(g) \rightarrow y\,SO_3(g)$

The equation will be balanced if
A $x=1$ and $y=1$
B $x=1$ and $y=2$
C $x=2$ and $y=1$
D $x=2$ and $y=2$
E $x=0$ and $y=0$

3.2 The correct representation for three molecules of nitrogen gas is
A $3N_2$ B 3N C N_3 D $2N_3$ E 6N

3.3 If a chromium ion is Cr^{3+} and a sulphide ion is S^{2-} the correct formula for chromium sulphide is
A CrS B Cr_3S_2 C Cr_2S D Cr_2S_3 E CrS_3

3.4 When the compound chromium(III) sulphide was analysed it was found to have 2 moles of chromium atoms combined with 3 moles of sulphur atoms. The correct equation for their reaction to form chromium(III) sulphide would be
A $Cr(s) + S(s) \rightarrow CrS(s)$
B $3Cr(s) + 2S(s) \rightarrow Cr_3S_2(s)$
C $2Cr(s) + 2S(s) \rightarrow Cr_2S_2(s)$
D $2Cr(s) + 3S(s) \rightarrow Cr_2S_3(s)$
E $3Cr(s) + S(s) \rightarrow Cr_3S(s)$

3.5 In one of the following lists all three elements exhibit two valencies, i.e. are all able to form two series of salts. They are:
A iron, copper, lead
B iron, copper, silver
C iron, silver, zinc
D copper, silver, gold
E copper, zinc, gold

3.6 In which one of the lists are all the formulae correct?
A CO_3, H_2SO_4, $CuCO_3$
B CH_4, H_2O, $NaCO_3$
C NH_4, HCl, $Al_2(SO_4)_3$
D SO_3, HNO_3, $Ca(OH)_2$
E HOH, HCO_3, NaOH

3.7 X, Y and Z (which are not their true symbols) are three elements. They each form compounds with hydrogen, having the formulae XH_4, HY, and ZH_3. The valency (combining capacity) of X, Y and Z are
A 1, 1, 3 B 2, 1, 3 C 3, 1, 3 D 4, 1, 3 E 4, 1, 2

3.8 The formula for phosphoric acid is H_3PO_4. The correct formula for each of the sodium salts it can form is
A $NaPO_4$, Na_2HPO_4, NaH_2PO_4
B Na_3PO_4, NaH_2PO_4, Na_2HPO_4
C $NaPO_4$, Na_2PO_4, Na_3PO_4
D $NaPO_4$, $Na(PO_4)_2$, $Na(PO_4)_3$
E $Na_3(PO_4)_3$, $Na_2(PO_4)_3$, $Na(PO_4)_3$

3.9 The metal M forms a chloride with the formula $MC1_2$. The formula for its sulphate is
A MSO_4 B M_2SO_4 C $M_2(SO_4)_3$ D $M(SO_4)_2$
E $M_3(SO_4)_2$

3.10 Which of the following facts cannot be obtained from a chemical equation?
A the reactants and resultants (products)
B the ratios of the reacting compounds
C the reaction conditions
D the physical state of the reactants and resultants (products)
E the names of all the reactants and resultants (products)

3.11 When copper(II) oxide is reduced by hydrogen the balanced symbolic equation is
A $Cu_2O(s) + H_2(g) \rightarrow 2Cu(s) + H_2O(l)$
B $CuO(s) + H_2(g) \rightarrow Cu(s) + H_2O(l)$
C $CuO(s) + 2H_2(g) \rightarrow Cu(s) + 2H_2O(l)$
D $2Cu_2O(s) + H_2(g) \rightarrow 2Cu(s) + 2H_2O(l)$
E $2CuO(s) + 2H_2(g) \rightarrow Cu(s) + H_2O(l)$

3.12 The correct ionic equation for the reaction between zinc and copper(II) sulphate solution is

A $Zn(s) + Cu_2SO_4(aq) \rightarrow ZnSO_4(aq) + 2Cu(s)$
B $Zn(s) + CuSO_4(aq) \rightarrow ZnSO_4(aq) + Cu(s)$
C $Zn(s) + Cu^{2+}(aq) \rightarrow Zn^{2}(aq) + Cu(s)$
D $Zn^{2+}(s) + Cu(aq) \rightarrow Zn(aq) + Cu^{2+}(s)$
E $Zn(s) + 2Cu^{2+}(aq) \rightarrow Zn^{2+}(aq) + 2Cu(s)$

Classification

Which of the following best fit the equations below? (You may use an answer once, more than once or not at all)

A 1 mole of solution +1 mole of solution → 1 mole of solid +2 moles of solution
B 1 mole of solution +1 mole of solution → 1 mole of solid +1 mole of solution
C 2 moles of solid +2 moles of solution → 2 moles of liquid +1 mole of gas
D 1 mole of solid +2 moles of solution → 1 mole of solution +1 mole of solution
E 1 mole of solid +2 moles of solution → 1 mole of solution +1 mole of gas

3.13 $2Na(s) + 2H_2O(l) \rightarrow 2NaOH(aq) + H_2(g)$

3.14 $Fe(s) + 2HCl(aq) \rightarrow FeCl_2(aq) + H_2(g)$

3.15 $BaCl_2(aq) + H_2SO_4(aq) \rightarrow BaSO_4(s) + 2HCl(aq)$

3.16 $CaCO_3(s) + 2HCl(aq) \rightarrow CaCl_2(aq) + H_2CO_3(aq)$
$(H_2O(l) + CO_2(g))$

3.17 $AgNO_3(aq) + HCl(aq) \rightarrow AgCl(s) + HNO_3(aq)$

Multiple completion

3.18 A metal X forms two ions X^{2+}, and X^{3+}, it will therefore form two compounds having the formulae

1 XSO_4 XCl_3
2 $X(OH)_2$ $X_2(SO_4)_3$
3 XCl XCl_3

3.19 $N_2(g) + 3H_2(g) \rightleftharpoons 2NH_3(g)$
The equation for the manufacture of ammonia using the Haber process is given above.

Which of the following can be deduced from the equation
1 The reaction is reversible
2 Ammonia is made by synthesis
3 Iron is used as a catalyst

Assertion–reason

3.20 $NaOH(aq) + H_2SO_4(aq) \rightarrow Na_2SO_4(aq) + H_2O(l)$

The balanced equation given above represents the neutralisation of sodium hydroxide solution by sulphuric acid solution **because** the correct equation for the reaction is

$2NaOH(aq) + H_2SO_4(aq) \rightarrow Na_2SO_4(aq) + 2H_2O(l)$

Chapter 4
Characteristics of Chemical Changes

Simple multiple choice

4.1 28 g of nitrogen contain the same number of moles as 32 g of
A hydrogen
B oxygen
C chlorine
D water
E sodium
(H = 1, O = 16, Cl = 35.5, Na = 23, N = 14)

4.2 Gay Lussac's law relating to gases refers to the
A volume of the gases that are reacting together
B volumes of the gases in relation to their temperature
C volumes of the gases in relation to their pressures
D kinetic energy of the reacting gases
E various densities of each reacting gas

4.3 Copper(II) oxide was made (i) by synthesis (ii) by thermal decomposition (iii) by precipitation of the hydroxide and drying. The copper oxide produced in each case had the formula CuO. These experiments could be used to prove the law of
A multiple proportions
B constant composition
C conservation of mass
D Avogadro
E the creation of matter

4.4 One mole of copper contains the same number of atoms as one mole of
A oxygen
B chlorine
C bromine
D neon
E hydrogen

Questions 4.5 to 4.8 refer to the compound sodium carbonate ($Na_2CO_3 10H_2O$) and anhydrous sodium carbonate (Na_2CO_3) (Na = 23, C = 12, O = 16, H = 1)

4.5 The molar mass of anhydrous sodium carbonate is
A 5.3 B 53 C 106 D 159 E 212

4.6 If one mole of sodium carbonate ($Na_2CO_3 10H_2O$) is heated to constant mass, the loss in mass would be
A 18 g B 26 g C 90 g D 180 g E 140 g

4.7 The percentage of water of crystallisation in washing soda ($Na_2CO_3 10H_2O$) is, to the nearest whole number
A 10% B 23% C 43% D 63% E 90%

4.8 To prepare 100 cm^3 of 0.1 M anhydrous sodium carbonate solution you would have to weigh out
A 0.23 g B 1.06 g C 5.1 g D 10 g E 10.6 g

4.9 If 2.88 g of copper(I) oxide was reduced using hydrogen the mass of copper remaining after the reaction had been completed is
(Cu = 64, O = 16)
A 0.32 g B 1.28 g C 1.56 g D 2.56 g E 2.88 g

4.10 When a fuel burns, energy is liberated and the reaction is said to be exothermic. Which of the graphs best show this kind of energy change?

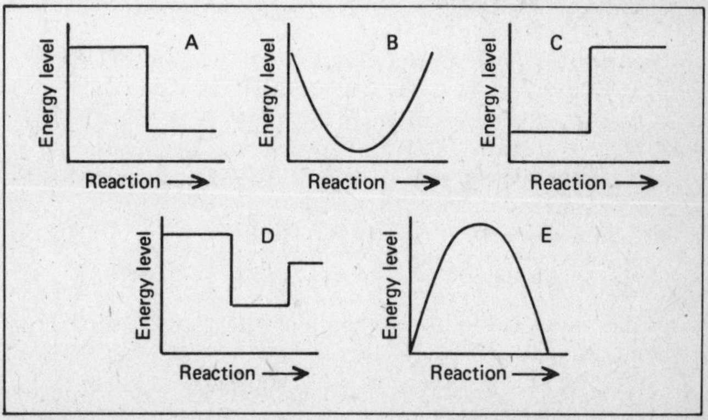

Figure 8

4.11 The main reason for using catalysts in industrial processes is to increase the
A amount of product formed
B temperature of the reaction
C rate of removal of the reactants
D rate of the removal of waste gases
E speed of the reactions

4.12 Various samples of propane gas (C_3H_8) were burned in air, and the respective masses of the carbon dioxide formed were recorded.

The results are given in the table. In which experiment is there a possible error in the mass of carbon dioxide produced?

Mass of propane	Mass of carbon dioxide formed
A 2.2 g	6.6 g
B 4.0 g	8.0 g
C 6.3 g	18.9 g
D 8.0 g	24.0 g
E 10.2 g	30. 6g

4.13 8 g of anhydrous copper(II) sulphate produced 12.5 g of copper(II) sulphate crystals. What is the formula of copper(II) sulphate crystals?
(Cu = 64, S = 32, O = 16)
A $CuSO_4$ B $CuSO_4.2H_2O$ C $CuSO_4.5H_2O$
D $CuSO_4.7H_2O$ E $CuSO_4.10H_2O$

4.14 What volume of 0.05 M hydrochloric acid is needed to react completely with 25 cm^3 of 0.15 M sodium carbonate solution?
A 60 cm^3 B 75 cm^3 C 100 cm^3
D 125 cm^3 E 150 cm^3

4.15 Which of the following contains the greatest number of moles?
(N = 14, O = 16, H = 1, Na = 23, Mg = 24)
A 40 g of nitrogen
B 40 g of oxygen

C 40 g of hydrogen
D 40 g of sodium
E 40 g of magnesium

4.16 Which of the following solutions contain the greatest number of moles?
A 50 cm^3 of M hydrochloric acid
B 50 cm^3 of 2 M sodium hydroxide
C 100 cm^3 of 0.5 M hydrochloric acid
D 50 cm^3 of 0.1 M sodium hydroxide
E 25 cm^3 of 5 M hydrochloric acid

4.17 A process which does not benefit from the use of a catalyst is the
A production of ammonia by the Haber process
B production of sodium chloride from sea water
C cracking of hydrocarbons
D action of hydrochloric acid on pure zinc
E production of sulphuric acid in the contact process

4.18 Le Chatelier's principle is concerned with the
A physical state of the chemicals in a reaction
B chemical changes occurring in a reaction
C equilibrium conditions that exist in a reaction
D volume of the gases produced in a reaction
E proportion of the elements in a reaction

4.19 The mass of one mole of iron(II) sulphate crystals ($FeSO_4.7H_2O$) is
(Fe = 56, S = 32, O = 16, H = 1)
A 1 g B 7 g C 152 g D 278 g E 305 g

4.20 What change will have no effect on the position of equilibrium of the reaction?
$N_2(g) + 3H_2(g) \rightleftharpoons 2NH_3(g) \qquad \Delta H = -46 kJ\,mol^{-1}$
A increasing the pressure in the reaction vessel
B decreasing the pressure in the reaction vessel
C adding a suitable catalyst
D increasing the temperature
E increasing the concentration of nitrogen and hydrogen

4.21 When 2 M hydrochloric acid is added to marble chips, which of the following conditions will not change the rate of the reaction?

A using twice as much 2 M HCl
B heating the solution
C powdering the marble beforehand
D doubling the concentration of the acid
E halving the concentration of the acid

The information in 4.22 also refers to 4.23.

4.22 27.40 g of an oxide of lead (A) is found on reduction to give 24.84 g of lead (O=16, Pb=207). The simplest empirical formula for the oxide (A) is

A Pb_3O_4 B PbO_2 C Pb_2O_3 D Pb_2O E Pb_4O_3

4.23 On reaction with nitric acid oxide (A) gives lead(II) nitrate, water and a second oxide of lead (B). The oxide (B) is found to contain 32 g of oxygen combined with the molar mass of lead. The simplest (empirical) formula for oxide B is

A Pb_3O_4 B PbO_2 C PbO D Pb_2O E Pb_2O_3

The information in 4.24 also refers to 4.25.

4.24 A sample of fertilizer containing ammonium sulphate, as the only nitrogen containing compound was found on analysis to contain 14% of nitrogen (H=1, N=14, O=16, S=32). The percentage of nitrogen in pure ammonium sulphate is to the nearest whole number

A 7% B 14% C 21% D 28% E 35%

4.25 The percentage of ammonium sulphate in the sample of fertilizer is to the nearest whole number

A 14% B 22% C 44% D 66% E 80%

4.26 Which of the following contain 6.02×10^{23} particles, i.e. Avogadro's number?

A 2 moles of copper
B $\frac{1}{2}$ mole of copper
C 1 mole of copper
D 3 moles of hydrogen
E 2 moles of hydrogen

Classification

Identify each of the following reactions as either a physical change (A) or chemical change (B)

4.27 Burning magnesium in oxygen
4.28 The rusting of iron
4.29 Melting flowers of sulphur
4.30 Dissolving sugar in tea
4.31 Making toffee from sugar
4.32 Oxygen being absorbed by the blood

Multiple Completion

4.33 2 g of hydrogen at stp
 1 has the same volume as 32 g of oxygen at stp
 2 occupies a volume of 22.4 dm^3
 3 will be given off from 1 mole of hydrochloric acid when reacted with zinc

4.34 71 g of chlorine and 64 g of copper
 (Cl = 35.5, Cu = 64)
 1 have the same volume
 2 contain the same number of particles
 3 contain a different number of particles

Assertion–reason

4.35 The action of heat on calcium carbonate is a reversible reaction **because** It is an example of thermal dissociation
4.36 The minimum energy needed for a reaction to proceed is called the activation energy **because** All reactions need energy to commence.

Structured question

4.37 The bromide of a metal M of formula MBr_2 reacts quantitatively with silver nitrate in accordance with the following equation

$$2AgNO_3(aq) + MBr_2(aq) \rightarrow 2AgBr(s) + M(NO_3)_2(aq)$$

In an experiment to determine the relative atomic mass of M, 4.6 g of the bromide were found to react with exactly 100 cm^3 of a solution of silver nitrate of concentration of 0.5 mol dm^{-3} (Br = 80)

(a) Calculate the number of moles of silver nitrate in $100\,cm^3$ of solution
(b) From the answer to (a) and the given equation calculate the number of moles of the bromide contained in 4.6 g of MBr_2
(c) From the answer to (b) calculate the molar mass of MBr_2
(d) Deduce the relative atomic mass of M.

Chapter 5
Atomic Structure and Bonding

Simple multiple choice

5.1 When sodium combines with chlorine, the following is true
A a gaseous compound is formed
B the product consists of positive and negative ions
C a chlorate is formed
D a covalent substance is formed
E a white crystalline solid with a low melting point is formed

5.2 All atoms of the same element have
A the same mass number (atomic mass)
B a different number of electrons
C the same number of protons
D different atomic numbers
E the same number of neutrons

5.3 When an atom of a non-metal becomes a negative ion with a single charge, the atom
A loses an electron
B shares an electron
C gains an electron
D gains a neutron
E loses a neutron

5.4 Which pairs of elements with the atomic numbers shown have similar properties
A 13, 22 B 3, 11 C 20, 32 D 6, 24 E 2, 4

5.5 Carbon is able to form many organic compounds by means of
A covalent bonds
B ionic bonds
C fermentation
D cracking
E homologous series

5.6 A compound was described as consisting of carbon and hydrogen only, in which all the carbon–carbon bonds are single covalencies. This is a description of

A a protein
B a detergent
C a carbohydrate
D an alkane
E an alcohol

5.7 The number of neutrons present in an atom of mass number (atomic mass) 133 and atomic number 55 is

A 55 B 78 C 133 D 188 E 200

5.8 The number of electrons in the outside shell of the most reactive metals is

A 1 B 2 C 3 D 4 E 7

5.9 When magnesium (atomic number 12) forms an ion, the electronic structure of the ion will be

A 2 B 2, 8 C 2, 8, 2 D 2, 8, 1 E 2, 8, 8

5.10 The relative atomic mass of chlorine (35.5) is not a whole number because

A the chlorine atom has 17 protons in the nucleus
B the chlorine atom has one more electron than it has protons
C the number of neutrons in the nucleus varies
D chlorine is a halogen
E it is the average mass of the protons neutrons and electrons

5.11 The nucleus of an atom contains

A neutrons
B protons
C electrons
D neutrons and protons
E neutrons and electrons

5.12 Sodium has an atomic number of 11, while oxygen has an atomic number of 8. The formula for sodium oxide is therefore

A NaO B NaO_2 C Na_2O D Na_2O_2 E Na_2O_3

5.13 Transition elements can be recognised because they have
A one unfilled shell of electrons
B two unfilled shells of electrons
C three unfilled shells of electrons
D no unfilled shells of electrons
E two shells of electrons

5.14 In the Periodic Table an element with an atomic number of 15 will be found in
A Group 5 Period 5
B Group 4 Period 3
C Group 5 Period 2
D Group 3 Period 3
E Group 5 Period 3

5.15 A particle containing 11 protons 10 electrons and 12 neutrons is
A a noble (inert) gas
B positively charged
C neutral
D negatively charged
E an atom

5.16 Two elements combine to form an ionic compound. This means that it
A conducts the electric current in the solid state
B is insoluble in water
C conducts electricity when fused (molten)
D is formed by sharing electrons
E has a low melting point

5.17 The description of a covalent compound is that it
A conducts electricity in any condition
B has a low melting point and boiling point
C is soluble in water
D is always a liquid or unstable solid
E contains charged particles

Classification

$^{40}_{19}A$ $^{32}_{16}B$ $^{23}_{11}C$ $^{35}_{16}D$ $^{14}_{7}E$

Which of these elements

5.18 are isotopes?
5.19 contain the greatest number of neutrons?

5.20 is in group 5 of the Periodic Table?
5.21 is the most reactive?
5.22 are in the same group but different period of the Periodic Table?
5.23 is most likely to be a non metal?
5.24 A A lattice of anions and cations. Non-conductor of the electric current.
B A lattice of cations held together in a mobile cloud of electrons. Conductor of the electric current.
C Atoms held in a giant covalent crystal. Non-conductor of the electric current.
D Atoms held in a giant covalent crystal. Conductor of the electric current.
E Closely packed molecules. Non conductor of the electric current.

Choose the best description for each of the following substances.
Copper, graphite, lead(II) bromide, magnesium oxide, silicon dioxide.

Multiple completion

5.25 Hydrogen, deuterium, and tritium are all isotopes of hydrogen because
1 they all have one electron
2 they are all gases
3 the all have one proton

5.26 Which of the following statements about atoms are correct?
1 The number of protons always equals the number of electrons
2 The number of neutrons always equals the number of electrons
3 The total number of positive charges on the neutrons always equals the number of negative charges on the electrons
4 the mass of the atom is mainly in the nucleus

Answer code
A 1 and 2
B 2 and 3
C 1 and 4
D 2 and 4
E 3 and 4

Assertion–reason

5.27 Sodium has a lower atomic number than potassium **because** One mole of sodium metal contains fewer atoms than one mole of potassium metal.

5.28 The nucleus of an atom consists of neutrons and protons **because** Neutrons are negatively charged and protons positively charged.

5.29 Diamond is an excellent example of a macromolecule **because** In diamond the units are bonded together covalently forming one molecule made up of a large number of atoms.

Structured question

5.30 The following sequence of elements, including their atomic numbers shows the third period of the Periodic Table

$_{11}Na$ $_{12}Mg$ $_{13}Al$ $_{14}Si$ $_{15}P$ $_{16}S$ $_{17}Cl$ $_{18}Ar$

(a) Give the electronic structure of (i) $_{11}Na$ (ii) $_{17}Cl$

(b) By means of a labelled diagram or otherwise, show the electronic changes that occur when sodium combines with chlorine to form sodium chloride and state the type of bonding involved.

(c) Explain why crystalline sodium chloride does not conduct electricity but fused (melted) sodium chloride does.

(d) Argon is a member of a group (family) of noble gases
- (i) Give the electronic structure of $_{18}Ar$
- (ii) For another element in the same group (family) as argon in the Periodic Table give its name and symbol.
- (iii) State what is special about the electronic structure of the noble gases.

(e) Nitrogen and phosphorus are members of the same group (family) of elements and show similar chemical properties
- (i) Give the electronic structure of $_{15}P$
- (ii) For the compound of phosphorus with hydrogen give its formula and state the type of bonding present.

(f) (i) Give the formulae of the oxides formed by magnesium and aluminium.
- (ii) To what class of oxides does each belong?

Chapter 6
Oxidation and Reduction

Simple multiple choice

6.1 A good reducing agent
- A is readily oxidised during a chemical reaction
- B contains hydrogen
- C is reduced during a chemical reaction
- D is usually a gas
- E must be combustible

6.2 Oxidation can best be defined as
- A the loss of oxygen from a substance
- B the loss of hydrogen from a substance
- C the loss of electrons from a substance
- D the gaining of electrons by a substance
- E the gaining of both oxygen and hydrogen by a substance

6.3 Which one of the following reactions is **not** an example of reduction?
- A $N_2(g) + 3H_2(g) \rightleftharpoons 2NH_3(g)$
- B $2SO_2(g) + O_2(g) \rightleftharpoons 2SO_3(g)$
- C $H_2S(g) + Cl_2(g) \rightarrow 2HCl(g) + S(s)$
- D $Mg(s) + 2HCl(aq) \rightarrow MgCl_2(aq) + H_2(g)$
- E $Mg(s) + O_2(g) \rightarrow 2MgO(s)$

6.4 Which of the following oxides are not reduced by hydrogen?
- A copper(II) oxide
- B magnesium oxide
- C iron(III) oxide
- D lead(II) oxide
- E tin oxide

6.5 Which of the following oxides can act as an oxidising agent?
- A carbon monoxide
- B calcium oxide
- C copper(I) oxide
- D lead(IV) oxide
- E sulphur dioxide

6.6 Old fashioned straw hats are bleached white by a process involving reduction. Eventually the hats start to go yellow again because
A the reduction process has continued
B we live in an oxidising atmosphere
C water vapour in the air reacts with the straw making it yellow
D the air contains gases that are reducing agents
E 80% of the air is nitrogen

6.7 When tap water is added to iron(II) sulphate the green solution formed soon changes to a yellow brown solution. The reason for this is
A the tap water has reacted with the iron(II) sulphate to form a brown insoluble compound
B the oxygen dissolved in the tap water oxidises the iron(II) sulphate solution to an iron(III) solution
C the water contains hydrogen which reduces the iron(II) sulphate solution to an iron(III) solution
D the iron(II) sulphate solution has gone rusty
E iron metal has been precipitated from the iron(II) sulphate solution

Multiple completion

6.8 Reducing agents commonly used in industry are:
1 carbon monoxide
2 chlorine
3 oxygen

6.9 In the electrolysis of lead(II) bromide, oxidation takes place
1 at the anode
2 at the cathode
3 in the electrolyte

6.10 $Cu^{2+}(aq) + H_2(g) \rightarrow Cu(s) + 2H^+(aq)$

This is called a redox reaction (equation) because
1 oxidation and reduction take place simultaneously
2 hydrogen acting as a reducing agent has itself been oxidised to water
3 copper(II) oxide acting as a reducing agent has itself been oxidised

Assertion–reason

6.11 $CuSO_4(aq) + Zn(s) \rightarrow ZnSO_4(aq) + Cu(s)$
In the above reaction copper(II) sulphate has been reduced to copper **because** Zinc like most metals is a good reducing agent.

6.12 When iron rusts it loses electrons **because** Oxygen has great affinity for iron.

Structured questions

6.13 (a) Define oxidation and reduction
(i) in terms of oxygen and hydrogen transfer
(ii) as it occurs in electron transfer reactions
(b) For each of the following reactions
(i) zinc and lead nitrate
(ii) copper(II) oxide and carbon monoxide
1 state the products
2 give an equation
3 state the conditions necessary for the reaction to occur
4 interpret the reaction in terms of the appropriate definitions of oxidation and reduction
(c) Compare the corrosion of iron and aluminium, stating why iron appears to be more easily oxidised than aluminium

6.14 (a) Define oxidation and reduction in terms of electron transfer
(b) The following ionic equations (A) and (B) involve magnesium and one of its compounds
A $Mg(s) + 2H^+(aq) \rightarrow Mg^{2+}(aq) + H_2(g)$
B $MgCO_3(s) + 2H^+(aq) \rightarrow Mg^{2+}(aq) + CO_2(g) + H_2O(l)$

State which one of the reactions (A) or (B) represents an oxidation and reduction and justify this in terms of your definitions in (a)
(c) In the electrolysis of fused (melted) lead(II) bromide, lead is formed at the cathode, and bromine at the anode. Write equations for the reactions occurring at the two electrodes and in each case state whether oxidation or reduction takes place.

Chapter 7
The Electrochemical (Activity) Series and Electrolysis

Simple multiple choice

7.1 Magnesium is a more reactive metal than lead. When it is placed into a solution of lead(II) nitrate

A magnesium nitrate is precipitated
B crystals of lead metal are formed
C effervescence occurs and hydrogen is evolved
D magnesium nitrate and lead(II) nitrate are formed
E there is no reaction

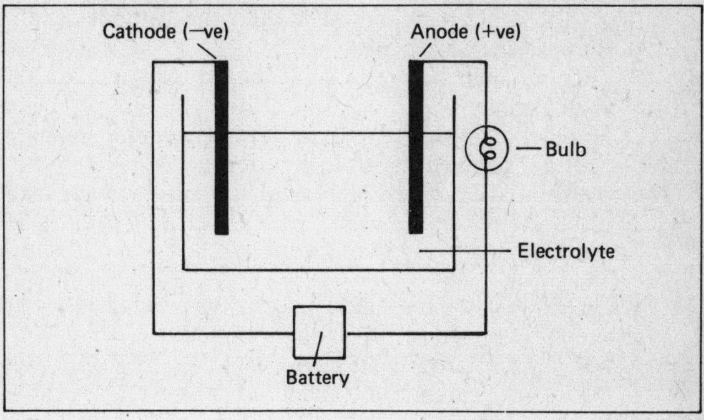

Figure 9

(This diagram refers to questions 7.2 to 7.5)

7.2 If the electrolyte used in the above apparatus was 100 cm^3 of 0.5 M sodium hydroxide solution, the volume of 2 M sulphuric acid needed to be mixed with the solution so that the bulb just started to glow was

A 25 cm^3 B 50 cm^3 C 100 cm^3 D 123 cm^3
E 150 cm^3

7.3 If in the same apparatus copper(II) sulphate was the electrolyte, and this was electrolysed using copper electrodes

A the electrolyte would change colour, going from blue to pale green
B oxygen would be formed at the cathode, and hydrogen at the anode
C the anode would get thicker, the cathode thinner and the blue solution would get paler
D the cathode would get thicker, the anode thinner and the blue solution would get paler
E copper would be deposited at the cathode and hydrogen would be formed at the anode

7.4 If sodium sulphate solution was electrolysed in the apparatus using platinum electrodes
A oxygen would be liberated at the cathode and hydrogen at the anode
B the solution around the anode would show a decrease in pH, while the solution around the cathode would show an increase in pH
C sodium would be deposited at the cathode and oxygen at the anode
D no reaction would take place because sodium sulphate does not conduct the electric current
E preferential discharge of sulphate ions would take place at the anode with hydrogen being liberated at the cathode

7.5 When an aqueous solution of hydrogen chloride was tested in the apparatus the bulb lit up and
A bubbles appeared at the cathode
B bubbles appeared at the anode
C bubbles appeared at both electrodes
D nothing else happened

7.6 Electrolysis can be best defined as
A the passage of current through an electrolyte without affecting the electrolyte
B the decomposition of an electrolyte by the passage of an electric current
C ionisation of the electrolyte
D a useful method of analysing covalent compounds
E a method of investigating the composition of an electrolyte

7.7 An unknown solution was electrolysed using platinum electrodes. The cathode became plated with a metallic

deposit, and a colourless gas was evolved at the anode. The solution was most likely to be
A lead(II) bromide
B copper(II) sulphate
C sodium sulphate
D sulphuric acid
E silver nitrate

7.8 In the electrolysis of sodium sulphate solution the reaction taking place at each of the electrodes is

	Anode	Cathode
A	$4H^+ + 4e^- \rightarrow 2H_2$	$4OH^- + 4e^- \rightarrow 2H_2O + O_2$
B	$2Na^+ + 2e^- \rightarrow 2Na$	$2H^+ + 2e^- \rightarrow H_2$
C	$4OH^- - 4e^- \rightarrow 2H_2O + O_2$	$4H^+ + 4e^- \rightarrow 2H_2$
D	$2H_2 + 2e^- \rightarrow H_2$	$2Na^+ + 2e^- \rightarrow 2Na$

7.9 Sodium metal can be produced on an industrial scale by the electrolysis of certain sodium compounds e.g. sodium hydroxide. Essential in this process is
A heat and an a.c. current
B heat and a d.c. current
C heat and the sodium hydroxide in solution
D a means of measuring the current flowing
E careful control on the concentration of the sodium hydroxide

7.10 Potassium is very high in the electrochemical (activity) series. This means that potassium
A is easy to extract from its compounds
B forms unstable compounds
C reacts slowly with solutions of other salts
D loses electrons easily
E gains electrons easily

7.11 In a series of experiments to investigate the reactivities of certain metals (labelled W, X, Y and Z) using displacement reactions the following results were obtained

W displaced Z
X displaced W
X displaced Z
Y displaced X

The order of reactivity, most reactive first is
A X, W, Y, Z B Y, X, Z, W C Z, W, X, Y
D Y, W, X, Z E Y, X, W, Z

7.12 As a result of space exploration a new metal oxide was discovered. Below hydrogen in the electrochemical (activity) series, a sample of the pure metal could be obtained by

A mixing the metal oxide with hydrogen
B placing a sample of the metal oxide into a gas jar of hydrogen
C heating the metal oxide in a porcelain crucible
D passing hydrogen over a heated sample of the oxide in a silica tube
E placing a sample of the oxide into a gas fired furnace

7.13 During the process of electrolysis there is no net change in electrical charge in the

A cathode
B anode
C electrolyte
D ions
E electrodes

7.14 The results of a series of investigations into the reactivities of three metals A, B, C and the gas hydrogen are given below

(i) C + A oxide $\rightarrow$ C oxide + A
(ii) B + dilute H_2SO_4 $\rightarrow$ no reaction
(iii) A + dilute H_2SO_4 $\rightarrow$ A sulphate + hydrogen
(iv) A + B sulphate $\rightarrow$ B + A sulphate

The order of reactivity of the metals and hydrogen, most reactive first is

A C, A, hydrogen, B,
B C, hydrogen, A, B,
C hydrogen, C, B, A,
D C, B, hydrogen, A,
E C, A, B, hydrogen.

7.15 The mass of the product (i) at the cathode and (ii) at the anode when 0.4 faraday of electricity passes through a solution of copper(II) sulphate using platinum electrodes is (Cu = 64, O = 16)

	(i) Cathode	(ii) Anode
A	6.4 g copper	1.6 g oxygen
B	64 g copper	16 g oxygen
C	12.8 g copper	3.2 g oxygen
D	128 g copper	32 g oxygen
E	19.2 g copper	4.8 g oxygen

7.16 Which element does not react with water or steam?
A calcium
B copper
C iron
D potassium
E zinc

7.17 During the electrolysis of water, hydroxyl ions are discharged at the anode. This discharge is best represented by
A $2H_2O + 2e^- \rightarrow 2H_2O + H_2$
B $4OH^- - 4e^- \rightarrow 2H_2O + O_2$
C $H_3O^+ + e^- \rightarrow H_2O + H$
D $4OH^- + 4e^- \rightarrow 2H_2O + O_2$
E $2OH^- - 2e^- \rightarrow H_2O + O_2$

7.18 Sodium chloride is an ionic compound, it will therefore
A dissolve readily in organic solvents
B melt very easily
C conduct electricity when solid
D form free ions in water
E consist of simple molecules

7.19 0.6 g of copper is deposited in a copper voltameter when a current of 1 A flows for $\frac{1}{2}$ hour. When $\frac{1}{2}$A of current flows for 2 hours the amount of copper deposited will be
(Cu = 64, 1F = 96 000 coulombs)
A 0.3 g B 0.4 g C 0.8 g D 1.2 g E 2.4 g

7.20 Which metal could **not** be formed by the electrolysis of aqueous solutions of its compounds using platinum electrodes?
A bromine
B chlorine
C copper
D sodium
E hydrogen

Classification
A zinc
B copper
C sodium
D aluminium
E silver

Which of the metals listed

7.21 is obtained on an industrial scale by the electrolysis of its naturally occurring oxide?

7.22 is nearest to iron in the electrochemical (activity) series and could be extracted in the blast furnace?

7.23 would be the easiest to obtain from its compounds?

7.24 is the most difficult to obtain from its compounds?

7.25 is the most likely to react with steam but not water?

7.26 is purified by electrolysis?

Multiple completion

7.27 In the electrolysis of copper(II) sulphate solution using platinum electrodes
1 reduction takes place at the cathode and oxidation at the anode
2 the cathode becomes coated with copper
3 the colour of the electrolyte is removed

7.28 When dilute sulphuric acid is electrolysed using platinum electrodes
1 oxygen and hydrogen are formed in the ratio 1:2
2 no metal above hydrogen in the electrochemical (activity) series could be used as electrodes
3 the concentration of SO_4^{2-} (sulphate) ions in the electrolyte would decrease.

Assertion–reason

7.29 The ease of decomposition of metal nitrates increases in descending order of the metal in the electrochemical (activity) series **because** All nitrates are soluble in water.

7.30 Solid sodium chloride does not undergo electrolysis **because** In the solid state the bonds between the ions are too strong to break down.

7.31 Tin is lower than iron in the electrochemical (activity) series, and is used to protect iron from corrosion by the process of tinning **because** Tin does not react with air, acids, or water.

Structured questions

7.32 In the table below are the reactions of metals A, B, C, D and E when placed in solutions of metal salts

Metal	Magnesium nitrate solution	Copper(II) sulphate solution	Iron(II) sulphate solution	Lead(II) nitrate solution
A	no reaction	reaction	no reaction	no reaction
B	no reaction	no reaction	no reaction	no reaction
C	no reaction	reaction	no reaction	reaction
D	no reaction	reaction	reaction	reaction
E	no reaction	reaction	reaction	reaction

When metal E is placed into a solution of a nitrate of D no reaction occurred

(a) Which metal in the table is most reactive?

(b) Which metal is the least reactive?

(c) State giving a reason for your answer which other named metal could **not** be used in this series of experiments.

(d) Place the metals in the table in order of decreasing reactivity.

(e) (i) If a mixture of the solutions used in the experiment in the table above were electrolysed in which order would the metal ions be discharged at the cathode?

(ii) Giving a reason for your answer state what would always be discharged at the anode.

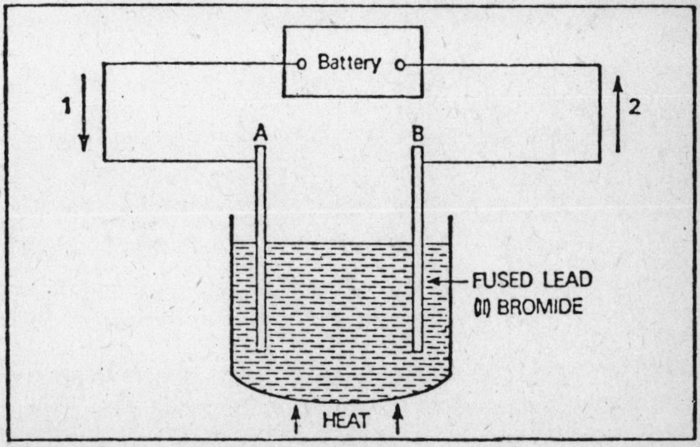

Figure 10

7.33 The diagram shows the electrolysis of fused lead(II) bromide A and B are carbon rods, the arrows showing the direction of the electron flow

(a) Which carbon rod is the cathode?

(b) Write symbols for the ions present in the electrolyte

(c) During the electrolysis which ions travel within the electrolyte to electrode B? Give your reason

(d) Write ionic equations for what happens at the electrodes during electrolysis

(e) What would be seen at the electrodes during electrolysis?

Chapter 8
Acids, Bases and Salts

Simple multiple choice

8.1 Acids:
A turn red indicator paper blue
B have a pH of 7 or more
C are often oxides of metals dissolved in water
D are often oxides of non metals dissolved in water
E sometimes contain hydrogen

8.2 The metals most commonly used to prepare hydrogen from a dilute acid are
A sodium, iron, zinc
B sodium, zinc, magnesium
C lead, tin, magnesium
D zinc, iron, magnesium
E iron, lead, zinc

8.3 A liquid with a pH of 8 is
A alkaline
B acidic
C neutral
D unreactive
E reactive

8.4 One oxide which dissolves in water to form an acid is
A calcium oxide
B vanadium(V) oxide
C sulphur dioxide
D dinitrogen oxide
E carbon monoxide

8.5 A colourless, odourless liquid which turns red indicator paper blue is
A ammonia solution
B ethanoic acid
C limewater solution
D water
E salt solution

8.6 Dilute hydrochloric acid was added to a white powder. The mixture frothed and bubbled. The white powder could have been
A copper(II) carbonate
B copper(II) chloride
C sodium carbonate
D sodium chloride
E sodium nitrate

8.7 The basicity of one molecule of any acid is the number of
A alkalis that would react with the acid
B hydrogen atoms it contains
C replaceable hydrogen atoms it contains

D replaceable oxygen atoms it contains
E salts it forms with any metal ion.

8.8 Which one of the following correctly completes the equation?

$$CaCO_3(s) + 2HCl(aq) \rightarrow$$

A $CaCl(s) + H_2O(l) + CO(g)$
B $CaCl_2(aq) + H_2O(l) + CO(g)$
C $CaCl_2(aq) + H_2O(l) + CO_2(g)$
D $CaCl(aq) + H_2O(l) + CO_2(g)$
E $CaCl_2(aq) + H_2(g) + H_2O(l) + CO_2(g)$

8.9 A acid + alkali
B acid + basic oxide
C acid + carbonate
D acid + metal
E synthesis of elements

Which method is the **most convenient** to prepare
(i) sodium hydrogensulphate?
(ii) iron(II) sulphide?
(iii) iron(II) sulphate?
(iv) sodium chloride?
(v) copper(II) carbonate?

8.10 The only soluble salt in the list is
A lead(II) nitrate
B lead(II) sulphate
C lead(II) chloride
D silver chloride
E lead(II) carbonate

8.11 The best method to use for the preparation of hydrogen chloride is to
A add dilute sulphuric acid to sodium chloride
B add dilute nitric acid to sodium chloride
C add concentrated sulphuric acid to sodium chloride
D electrolyse sodium chloride solution
E react hydrogen with chlorine

8.12 When dry hydrogen chloride gas was dissolved in dry methylbenzene (toluene) the resultant solution had no effect on pH paper, because
A toluene is not an acid
B hydrogen chloride has no effect on pH paper

C hydrogen chloride contains water
D acids need water to ionise
E toluene is an alkali and neutralises hydrogen chloride

8.13 After burning magnesium in oxygen, the pH of the solution of the oxide with water is
A 1 B 3 C 5 D 7 E 9

8.14 A white salt is heated in a crucible and *vapourises* completely leaving no residue in the crucible. The solid is
A calcium carbonate B sodium chloride
C sodium carbonate D ammonium chloride
E zinc oxide

8.15 The process taking place in 8.14 above is called
A crystallisation B decomposition
C diffusion D distillation
E sublimation

8.16 Epsom salts has the formula $MgSO_4.7H_2O$. What mass of water is lost when 1 mole of Epsom salts is dehydrated (Mg=24, S=32, O=16, H=1)?
A 5g B 18g C 90g D 126g E 210g

8.17 The pH of a liquid is the
A hydrogen ion concentration of the liquid
B number of replaceable hydrogen ions in the liquid
C number of hydrogen ions present in one molecule of the liquid
D basicity of the liquid
E number of oxygen ions in the liquid

Classification

Five solutions have the following pH values (referring to questions 8.18 to 8.22)
A pH=1
B pH=3
C pH=6.5
D pH=8
E pH=11
Which solution is
8.18 the strongest alkali?
8.19 the strongest acid?
8.20 nearest to neutral?

8.21 Which two solutions when mixed together could produce a neutral solution?

8.22 Which of the solutions would you use to neutralise an alkali spillage on the skin?

A sodium sulphate (Na_2SO_4)
B sulphuric acid (H_2SO_4)
C sodium carbonate (Na_2CO_3)
D potassium hydroxide (KOH)
E calcium hydroxide ($Ca(OH)_2$)

Which one of the above solutions

8.23 would you use to neutralise an acid spillage on clothing?

8.24 produces a white precipitate with barium chloride solution?

8.25 reacts with magnesium to produce magnesium sulphate and hydrogen?

Structured questions

8.26 In the preparation of sodium sulphate crystals 25 cm^3 of 2M sodium hydroxide solution was reacted with 2M sulphuric acid in the presence of suitable indicator until the indicator just changed colour. The solution was evaporated to half its volume and allowed to cool and crystallise out. The mother liquor was poured away, the crystals dried between filter papers and stored.

(a) Name a suitable indicator for this experiment
(b) Write a balanced equation for the reaction
(c) How many moles of sodium hydroxide are there in 25 cm^3 of 2M sulphuric acid?
(d) By reference to your equation state how much 2M sulphuric acid you would expect to add to neutralise the sodium hydroxide.
(e) What do you think is wrong in the preparation of sodium sulphate in the account given above?

8.27 In the preparation of zinc sulphate an indicator is not required. The equation for this preparation is

$$H_2SO_4(aq) + ZnCO_3(s) \rightarrow ZnSO_4(aq) + H_2O(l) + CO_2(g)$$

(a) What is observed when the zinc carbonate is added to the acid?
(b) In this preparation the solution obtained after the reaction was filtered. Give a reason for this.
(c) What mass of zinc carbonate would you expect to use to neutralise 25 cm^3 of 2M sulphuric acid solution?

(d) The zinc sulphate solution obtained in the experiment was used to prepare a sample of lead sulphate, an insoluble salt. The equation is as follows

$$ZnSO_4(aq) + Pb(NO_3)_2(aq) \rightarrow PbSO_4(s) + Zn(NO_3)_2(aq)$$

(i) How would you ensure that all the lead sulphate had been precipitated?
(ii) How would you separate the lead sulphate from the solution?
(iii) Why do you think lead sulphate is washed repeatedly after separation?
(iv) Give the name of another solution you could have used other than zinc sulphate.

Chapter 9

The Periodic Classification of the Elements

Simple multiple choice

9.1 Elements in the Periodic Table are arranged in order of their

A atomic masses
B atomic numbers
C reactivities
D densities
E number of neutrons

9.2 The vertical rows in the Periodic Table are called

A periods
B lists
C groups
D triads
E families

9.3 Which of the following is **not** a property of the alkali metals?

A they all have one electron on their outside shell
B they all form ions of the A^+ type
C they can all be extracted by the electrolysis of solutions of their chlorides
D they all react violently with cold water
E they all form alkalis

9.4 All the elements in the same group of the Periodic Table will have the same

A melting points
B densities
C molar masses
D reactions with oxygen
E colours

9.5 The scientist responsible for producing the modern version of the Periodic Table was

A Newlands
B Dobereiner
C Newton
D Mendeleeff
E Dalton

9.6 The electron arrangement of an element X is 2, 8, 2. This means that the element can be found in

A Group 8 Period 2
B Group 2 Period 8

C Group 2 Period 4
D Group 2 Period 3
E Group 2 Period 10

9.7 Rubidium is in the same family of elements as sodium and potassium. This means that rubidium
A forms an insoluble oxide
B reacts with a halogen to give a solid halide
C is stored under water
D is a hard shiny metal
E forms an acidic oxide

9.8 Which pair of elements will react most violently together?
A hydrogen and fluorine
B sodium and chlorine
C potassium and bromine
D caesium and fluorine
E potassium and fluorine

9.9 The noble (inert) gases are placed in the same group of the Periodic Table because
A they are all gases
B they are all unreactive gases
C they all have eight electrons in their outside shells
D they are all non-metals
E they all react with gases in the air.

9.10 Which of the following contain elements with two unfilled shells of electrons?
A simple elements
B transition elements
C rare earths
D inert gases
E halogens

9.11 Lithium is found in Group 1 of the Periodic Table, and oxygen is in Group 6. The simplest formula for lithium oxide is
A LiO B LiO_2 C Li_2O D Li_2O_2 E Li_2O_6

Classification

The diagram on the next page is part of the Periodic Table of the elements. The letters indicate the positions of some of the elements, these letters are **not** the symbols for the elements.

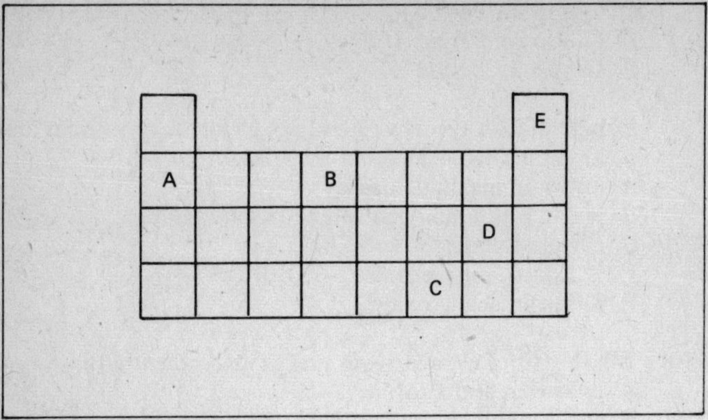

Figure 11

(The diagram refers to questions 9.12 to 9.16).

Which letter represents

9.12 a halogen?
9.13 an element that reacts violently with water?
9.14 a gas?
9.15 a transition element?
9.16 an element that forms covalent compounds?

Multiple completion

9.17 In going from fluorine to astatine in Group 7 the
1 colour of the elements darken
2 reactivity decreases
3 electropositive nature increases.

9.18 The noble (inert) gases are
1 unreactive
2 obtained from the air
3 coloured

9.19 The correct formula for compounds of Group 1 metals is
1 M_2O
2 MCl
3 $M(OH)_2$

Assertion–reason

9.20 The elements carbon and silicon can both form long chain compounds **because** Both are in Group IV of the Periodic Table

9.21 Sodium has a lower atomic number than potassium **because** Potassium is more reactive than sodium

9.22 Iodine can be purified by sublimation **because** It is a halogen

Structured question

9.23 Part of the Periodic Table of the elements is given below

GROUP

	1	2											3	4	5	6	7	8
1	1 H																	2 He
			TRANSITION ELEMENTS															
2	3 Li	4 Be	21 ↓	22 ↓	23 ↓	24 ↓	25 ↓	26 ↓	27 ↓	28 ↓	29 ↓		5 B	6 C	7 N	8 O	9 F	10 Ne
3	11 Na	12 Mg	↓	↓	↓	↓	↓	↓	↓	↓	↓		13 Al	14 Si	15 P	16 S	17 Cl	18 Ar
4	19 K	20 Ca	21 Sc	22 Ti	23 V	24 Cr	25 Mn	26 Fe	27 Co	28 Ni	29 Cu	30 Zn	31 Ga	32 Ge	33 As	34 Se	35 Br	36 Kr

PERIOD

Figure 12

(a) The name and symbol of one element which reacts with water to form hydrogen is...

(b) The name and symbol of one element which does not react with hydrogen or oxygen is...

(c) Some elements react with oxygen to form a gas which dissolves in water to make an acid. One such element is...

(d) Find the element manganese (Mn) and write down two properties you would expect this element or its compounds to have.

(e) The element lithium (Li) forms lithium chloride with chlorine. Write down the following details you would expect of this compound.

(i) formula

(ii) colour of the solid

(iii) solubility or not in water

(iv) electrical conductivity or not of the molten compound

(f) Arsenic (As) forms a gaseous compound with hydrogen. A probable formula for this gas is...

Chapter 10
Metals and their Compounds

Simple multiple choice

10.1 Which of the following is not a metal?
A calcium   B sodium
C phosphorus   D chromium
E copper

10.2 Which of the following is not an alloy?
A titanium   B bronze
C stainless steel   D brass
E solder

10.3 If calcium lies between sodium and aluminium in the electrochemical (activity) series of metals, the most likely method of extracting the metal calcium is
A heating calcium oxide strongly
B heating a mixture of calcium oxide and zinc
C adding sodium to calcium chloride solution
D electrolysis of calcium chloride solution
E electrolysis of molten calcium chloride

10.4 Which of the following compounds is a base?
A calcium oxide
B lead(II) nitrate
C sodium chloride
D sulphur dioxide
E copper(II) sulphate

10.5 The conversion of cast iron into mild steel is essentially a process of
A decomposition   B electrolysis
C purification   D reduction
E synthesis

10.6 The mass of aluminium in tonnes obtained from the complete electrolysis of 10.2 tonnes of aluminium oxide is ($Al=27$, $O=16$)
A 2.7   B 5.1   C 5.4   D 13.5   E 20.4

10.7 Which of the following metals will not react with dilute sulphuric acid to produce hydrogen gas?
A aluminium
B copper
C iron
D magnesium
E zinc

10.8 In the extraction of sodium in the Downs process strontium chloride is mixed with the sodium chloride to
A increase the viscosity of the molten sodium chloride
B improve the conduction of the mixture
C lower the melting point
D decrease the corrosion of the iron cathode
E help the movement of the ions in the process

10.9 An efflorescent compound
A gains water from the atmosphere and dissolves in it
B is a good drying agent
C loses water of crystallisation to the air
D crumbles when exposed to dry air
E always becomes hydrated when exposed to the air

10.10 A solid substance X has the following physical properties
(i) it is a white crystalline solid
(ii) it has a high melting point
(iii) it is a good conductor in the molten state
The solid is
A magnesium oxide
B sodium chloride
C lead(II) sulphate
D copper(II) sulphate
E sulphur(VI) oxide

10.11 Which one of the following is an alkali metal?
A aluminium
B copper
C lead
D zinc
E sodium

10.12 A metal Z forms a chloride of formula ZCl. When treated in a flame test a lilac flame was produced. Z is
A calcium
B copper
C sodium
D potassium
E iron

10.13 When the metal aluminium is anodised the
A metal is used in an electrolytic cell
B oxide layer covering the metal is thickened
C metal becomes more reactive
D conductivity of the metal increases
E metal is purified

10.14 A metal high in the electrochemical (activity) series
A can be extracted from its ore by heating it with carbon
B reacts readily with cold water to give an alkali and hydrogen
C burns to form an oxide with a pH less than 7
D forms unstable compounds
E is an oxidising agent

10.15 An ore is
A another form of a pure metal
B a metal oxide
C a substance from which a metal can be extracted economically.
D a stable form of a metal alloy
E a substance from which a metal can be extracted by heating with carbon

10.16 Which is not typical of a metal?
A it is a reducing agent
B it has three or less electrons on its outside shell
C it forms negative ions
D it forms basic oxides
E some can form alkali solutions

10.17 Many metals can be hammered into sheets and drawn into wires, and are therefore referred to as being malleable and ductile. This is because
A the metal lattice is easy to disrupt
B the bonds between the atoms are very strong
C the layers of atoms slip easily over each other
D the metal atoms are strongly attracted to each other
E the metal atoms are arranged in chains

10.18 Metals are good electrical conductors because
A they contain free electrons
B they are silvery in colour and attract radiation
C their particles are very closely packed
D they are also good conductors of heat
E they have high tensile strengths

10.19 Sodium and potassium compounds behave in similar ways because
A they have the same density
B they both occur as chlorides
C they are extracted in the same way
D they have similar outer electron structures
E they both have two electrons on the first shells

10.20 Farmers lime their fields because it
A neutralises acid soils
B neutralises alkali soils
C gives nitrogen to the soil
D assists in the process of photosynthesis
E is an important and effective insecticide

10.21 In the production of aluminium the carbon anodes have to be replaced periodically because
A they become brittle because of the intense heat in the cell
B their conductivity decreases as time goes on
C they burn in the oxygen liberated at them
D they are corroded away by the environment in the cell
E they become coated with aluminium and are no longer able to conduct effectively

10.22 In aluminium extraction, cryolite is mixed with alumina in the cell feed because it
A lowers the melting point of the alumina and decreases the fluidity of the cell feed
B increases the melting point of the aluminium oxide and decreases the fluidity of the cell feed
C increases the melting point of the alumina and increases the fluidity of the cell feed
D decreases the melting point and increases the fluidity of the cell feed
E decreases the melting point and has no effect on the fluidity of the cell feed

10.23 In the purification of bauxite (impure aluminium oxide) in the Bayer process, use is made of the
A solubility of aluminium oxide in water
B amphoteric nature of the aluminium oxide
C fact that bauxite is a basic oxide
D fact that sodium hydroxide is a very strong base
E fact that sodium hydroxide dissolves aluminium oxide and all the impurities

10.24 Which one of the following statements about alkali metals is untrue?
A their oxides are amphoteric
B they displace hydrogen from water
C they are very reactive
D they form solid hydrogen carbonates
E their compounds are generally very stable

10.25 A major use of sodium nitrate is in
A gun powder
B baking powder
C fuses for explosives
D preservatives
E making hydrochloric acid

10.26 All acids give carbon dioxide with calcium carbonate except
A hydrochloric acid
B ethanoic acid (vinegar)
C phosphoric acid
D sulphuric acid
E nitric acid

10.27 Sodium hydrogencarbonate ($NaHCO_3$) is used as a raising agent in self raising flour because it
A is a light airy substance
B mixes well with flour
C releases hydrogen when heated
D releases carbon dioxide when heated
E neutralises acidic compounds contained in the flour

10.28 A substance which is not a form of calcium carbonate is
A quicklime
B limestone
C calcite
D chalk
E marble

10.29 Which of the following nitrates does not break down when heated to give an oxide, nitrogen dioxide and oxygen?
A zinc nitrate
B potassium nitrate
C lead(II) nitrate
D copper(II) nitrate
E iron(III) nitrate

10.30 The best description of the reaction that occurs when water is added to calcium oxide (quicklime CaO) is
A it dissolves readily, and great heat is evolved
B it sizzles, swells up, and great heat is evolved
C it decomposes readily with the formation of a solution of calcium hydroxide

D it forms a white slurry which is used to whitewash walls
E a very slow reaction occurs which is aided by heating

10.31 A lime kiln produces
A lime and carbon dioxide
B quicklime and carbon dioxide
C calcium carbonate and carbon dioxide
D limestone and carbon dioxide
E limewater solution

10.32 Blue copper(II) sulphate crystals are warmed in a test tube. The crystals will
A turn white and a vapour is evolved
B melt to give a blue solution
C turn pink
D break down to give copper(II) oxide and sulphur dioxide
E change directly into metallic copper

10.33 The element tin is extracted from its oxide by strongly heating the oxide with powdered coal. Coal is used because it
A removes impurities
B removes impurities and provides sufficient heat for the reaction to proceed
C burns readily to form reducing gases
D contains carbon which reduces the oxide
E is lower than tin in the electrochemical (activity) series

10.34 Sodium hydrogencarbonate ($NaHCO_3$) and sodium carbonate (Na_2CO_3) can be distinguished from each other by
A heating and testing any gas evolved
B their physical properties
C their colours
D their reaction with acids
E their reaction with alkalis

10.35 Aluminium reacts slowly with dilute acids because
A it is low in the electrochemical (activity) series
B it only reacts with concentrated acids
C the salts of aluminium are insoluble
D it is covered with a coating of oxide which is unreactive

E the reaction needs a catalyst to increase the speed of the reaction

10.36 Hot metal + steam → metal oxide + hydrogen
Which of the following would you use to carry out the above reaction?
A copper
B iron
C zinc
D lead
E carbon

10.37 Which oxide will not be reduced by heating with powdered carbon over a bunsen burner?
A mercury(II) oxide
B lead(II) oxide
C magnesium oxide
D iron(III) oxide
E copper(II) oxide

10.38 In some experiments involving magnesium, pupils are advised to clean the metal with emery cloth before the experiment. This is done to
A provide a good even surface for the reaction
B remove the oxide layer formed by the air
C roughen up the surface to provide more area of contact
D ensure that the metal is free from moisture or oil
E make sure that both sides of the magnesium are the same

10.39 Iron is most likely to go rusty if
A the iron is put into tap water
B the iron is put into distilled water
C the iron is heated in air
D the iron is placed in boiled water
E it is surrounded by dry air

10.40 Magnesium metal will react when strongly heated in each of the following gases with the exception of
A oxygen
B nitrogen
C neon
D sulphur dioxide
E carbon dioxide

Classification

10.41 Black, green, yellow, blue, grey. Choose the correct colour for each of the following
(i) calcium carbonate

(ii) iodine
(iii) copper(II) sulphate
(iv) iron(II) sulphate
(v) silver iodide

10.42 Aluminium, carbon, copper, lead, zinc. Which metal is
(i) obtained by electrolysis
(ii) used in the alloy solder
(iii) used to galvanise iron
(iv) refined by electrolysis after it has been extracted from its ore
(v) used on a large scale as a reducing agent.

10.43 Calcium, copper, lead, sodium, zinc. Choose the metal
(i) whose carbonate and hydroxide are both soluble
(ii) whose anhydrous chloride is a drying agent
(iii) that is least electro positive
(iv) that lies between aluminium and iron in the electrochemical (activity) series
(v) that forms an oxide which oxidises hydrogen chloride to chlorine

Multiple completion

10.44 The metal that displaces copper from copper(II) sulphate is
1 zinc
2 iron
3 silver

10.45 The compound that decomposes to form an oxide is
1 copper(II) nitrate
2 lead(II) nitrate
3 sodium nitrite

Assertion–reason

10.46 When the coating of tin on steel is broken the iron rusts more quickly than normally **because** Tin is lower down the electrochemical (activity) series than iron

Structured questions

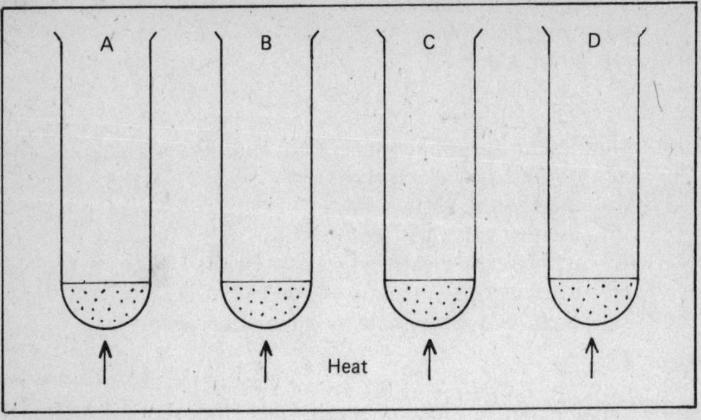

Figure 13

10.47 Each of the test tubes A, B, C and D contain exactly 1.24 g of copper(II) carbonate. When this compound is heated the following reaction occurs

$$CuCO_3(s) \rightarrow CuO(s) + CO_2(g)$$

Each tube was heated, cooled in a desiccator and weighed. Each tube was heated four times. The results obtained are given in the table

Sample	Mass after each heating/g			
	1	2	3	4
A	1.30	0.92	0.93	0.92
B	0.92	0.91	0.82	0.80
C	0.84	0.82	0.80	0.80
D	0.80	0.80	0.80	0.80

(a) In which tube has most copper(II) carbonate decomposed after the first heating?

(b) Which set of results need checking for error? Give a reason for your answer?

(c) Why is there no change in mass in results 3 and 4 in the case of sample A?

(d) Using the table say how much copper(II) oxide is obtained from 2.48 g of copper(II) carbonate

(e) Using the equation check your answer to part (d) by actually calculating the amount of copper(II) oxide obtained from 2.48 g of copper(II) carbonate

(Cu = 64, C = 12, O = 16)

10.48 Sodium hydroxide, chlorine, and a gas which is lighter than air are produced when brine is electrolysed. The anode in the cell is made of graphite

(a) (i) Name the gas which is lighter than air, produced in this process

(ii) Name the ions present in the brine

(iii) Why is it better to use a graphite anode rather than one made of metal?

(iv) Suggest a part of the country which would be best suited for this process to be sited and carried out on a large scale. Give a reason for your answer.

(b) Name a metal which is

(i) purified by electrolysis

(ii) extracted by electrolysis

10.49 The corrosion of iron was investigated by giving six identical iron nails different treatments. One other nail was left untreated. All seven nails were then left for several weeks exposed to the atmosphere. The results are given in the table

Nail	Treatment	Cost of treatment	Mass of nail and coating before exposure to the atmosphere g	Mass of nail and coating after exposure to the atmosphere g
A	Waxed	Low	5.0	5.3
B	Oiled	Low	6.0	5.2
C	Chromium plated	High	6.0	6.0
D	Painted	Low	6.0	6.4
E	Untreated		5.9	7.1
F	Galvanised	Fairly High	6.0	6.1
G	Dipped in salt solution	Low	6.0	6.7

(a) Name **two** naturally occurring substances which must be present for iron to rust
(b) Which nail was **best protected** against corrosion?
(c) Which nail received a treatment which made the corrosion worse than it would have been if left untreated?
(d) Which nail received a treatment which is usually used on car bumpers?
(e) Which nail received a treatment which is usually used on iron railings?
(f) In which case was there an obvious mistake in the weighing of the nail and the coating after the experiment?
(g) When iron rusts iron(II) oxide is one of the products. Name the chemical process that has taken place.
(h) What is the mass of one mole of iron(II) oxide? (Fe = 56, O = 16)

10.50 (a) State what is **seen** in **each** of the following series of reactions:
(i) Equal masses of lead foil are added to equal volumes of separate cold solutions of sodium nitrate, zinc(II) nitrate, and copper(II) nitrate, **each** having the same molar concentration.
(ii) The same numbers of moles of zinc(II) carbonate, anhydrous sodium carbonate and copper(II) carbonate are separately heated under the **same** conditions for the **same** time. The volume of dry gas evolved was measured in **each** case.
(iii) Zinc and copper foil are added separately to dilute hydrochloric acid, containing a few drops of a solution of a copper(II) salt, under the same conditions.

From the results place the elements copper, lead, sodium and zinc in their correct order in the electrochemical (activity) series, justifying your order by giving explanations for **each** series of reactions.

(b) Describe and explain what happens when the following metallic salts are heated:
(i) iron(II) sulphate heptahydrate ($FeSO_4.7H_2O$);
(ii) copper(II) sulphate pentahydrate ($CuSO_4.5H_2O$);
(iii) sodium nitrate.

Chapter 11
Non-metals and their Compounds

Simple multiple choice

11.1 The gas found in the greatest quantity in air is
A nitrogen B oxygen
C argon D hydrogen
E carbon dioxide

11.2 The catalyst used in the preparation of oxygen by the decomposition of hydrogen peroxide is
A platinum
B magnesium oxide
C manganese(IV) oxide
D manganese(II) oxide
E magnesium

11.3 When a piece of phosphorus was burned in a fixed volume of air, it eventually went out. The volume of air remaining behind was
A 100% B nil C over 75% D under 50%
E under 30%

11.4 The gas not normally found in air is
A nitrogen B oxygen
C argon D hydrogen
E carbon dioxide

11.5 A property typical of non metals is that they
A have high melting points
B form covalent hydrides
C form ionic chlorides
D form oxides that are basic
E are very soluble in water

11.6 Sulphur is extracted from beneath the earth using the Frasch process. This process depends on the
A colour of sulphur
B density of sulphur
C melting point of sulphur
D reactivity of sulphur
E allotropy of sulphur

11.7 The compound that is present in water giving temporary hardness is
A calcium carbonate
B calcium sulphate
C calcium hydrogencarbonate
D magnesium carbonate
E sodium carbonate

11.8 The gas normally used to replace nitrogen in the artificial atmosphere used in deep sea diving is
A argon
B carbon dioxide
C hydrogen
D helium
E neon

11.9 The substance **not** used in the contact process for the production of sulphuric acid is
A oxygen
B sulphur
C water
D vanadium(V) oxide (vanadium pentoxide)
E hydrogen

11.10 A colourless gas that dissolves in water to form a solution with a pH of 11 could be
A oxygen
B ammonia
C chlorine
D sulphur dioxide
E carbon dioxide

11.11 The reaction in which sulphuric acid is acting as an oxidising agent is
A $H_2SO_4(aq) + S(s) \rightarrow H_2O(l) + 2SO_2(g)$
B $H_2SO_4(l) + KCl(s) \rightarrow KHSO_4(aq) + HCl(g)$
C $2H_2SO_4(l) + S(s) \rightarrow 2H_2O(l) + 3SO_2(g)$
D $H_2SO_4(aq) + NaOH(aq) \rightarrow NaHSO_4(aq) + H_2O(l)$
E $H_2SO_4(aq) + NaOH(aq) \rightarrow Na_2SO_4(aq) + H_2O(l)$

11.12 The best description of a neutral oxide is:
A a solid insoluble in water but soluble in dilute hydrochloric acid
B a liquid with a pH of 6.5
C a gas that dissolves in water to form a solution which has no effect on Universal Indicator
D a liquid that produces carbon dioxide with carbonates
E a solid that decomposes with heat

11.13 The statement which is wrong about nitric acid is that it is
A a nitrate
B an oxidising agent
C a dibasic acid
D an acid that displaces carbon dioxide from carbonates
E a compound that contains hydrogen

11.14 The process that does not soften permanent hard water is
A boiling
B adding soap
C using ion exchange compounds
D adding excess sodium carbonate
E distillation

11.15 Which of the following oxides does not produce an acid when dissolved in water?
A sodium oxide
B sulphur dioxide
C dinitrogen pentoxide
D carbon dioxide
E phosphorus(V) oxide

11.16 A company offered the following fertilisers to a farmer needing to add nitrogen to his soil. Which one should he choose to provide the maximum amount of nitrogen?

Fertilizer	Mass of nitrogen per mole	Mass of 1 mole
A	28	80
B	14	101
C	28	132
D	14	85
E	28	60

11.17 The substance that needs oxygen in its manufacture is
A ammonium sulphate B nylon
C polystyrene D steel
E lime

11.18 A yellow powder when heated in air burned with a blue flame and produced a gas with a pH of 2. The gas also

bleached moist litmus paper. The gas was
A chlorine
B sulphur dioxide
C sulphur(VI) oxide (sulphur trioxide)
D carbon dioxide
E nitrogen dioxide

11.19 Pure water is a poor conductor because it
A contains no ions
B contains a few ions
C contains a large number of ions
D is a very good solvent
E is a liquid

11.20 A fertiliser was heated with sodium hydroxide solution and the gas evolved changed red litmus paper blue. This proves that the fertiliser contains
A carbon
B phosphorus
C potassium
D nitrogen
E sulphur

11.21 In which reaction is hydrogen **not** produced?
A heating magnesium in steam
B electrolysing an aqueous solution of sodium sulphate
C reacting copper with concentrated sulphuric acid
D reacting sodium hydride with water
E adding sodium metal to cold water

11.22 Hydrogen can be collected by the downward displacement of air because it is
A more dense than air
B as dense as air
C less dense than air
D part of an explosive mixture with air
E insoluble in cold water

11.23 An all glass apparatus is needed in the preparation of nitric acid because it is
A a powerful reducing agent
B a much cleaner method
C a powerful oxidising agent
D a soluble nitrate
E reactive with all silicates

11.24 A substance was heated very strongly in air, and the acid gas sulphur dioxide was formed. This means that the substance heated must have

A been sulphur
B been a sulphate
C been a sulphite
D contained sulphur
E contained sulphur dioxide

11.25 Which of the following would be the least likely to produce oxygen?

A heating mercury(II) oxide
B removal of nitrogen from the air
C heating red lead oxide
D production of zinc from zinc oxide
E heating hydrogen peroxide and manganese(IV) oxide

11.26 Fur in kettles is due to

A water standing in the kettles overnight
B the decomposition of permanently hard water
C the decomposition of temporary hard water
D water evaporating completely from the kettle
E a chemical reaction between the metal in the kettle and the water

11.27 Ammonia is collected by the downward displacement of air because it is

A very soluble in water
B more dense than air
C the same density as air
D less dense than air
E a reactive gas when mixed with air

11.28 When sulphuric acid is produced in the contact process, the sulphur(VI) oxide (sulphur trioxide) formed is

A dissolved in water
B dissolved in weak acid
C dissolved in strong acid
D liquefied, pressurised and stored in an aerosol for use
E allowed to solidify

11.29 One of these statements about the preparation of dry chlorine is incorrect

A the reaction must be heated
B manganese(IV) oxide catalyses the reaction
C the gas is passed through concentrated sulphuric acid
D chlorine is collected by the upward displacement of air

E the gas is passed through water to free it from acid spray

11.30 Chlorine is prepared by
A the action of heat on a chlorate
B heating a mixture of concentrated sulphuric acid and common salt
C oxidising hydrochloric acid
D electrolysing sodium chloride solution
E mixing sodium chloride and an oxidising agent

11.31 Which equation represents nitric acid acting as an oxidising agent?
A $H_2SO_3(aq) + 2HNO_3(l) \rightarrow H_2SO_4(aq) + H_2O(l) + 2NO_2(g)$
B $Na_2CO_3(s) + HNO_3(aq) \rightarrow NaNO_3(aq) + H_2O(l) + CO_2(g)$
C $NaOH(aq) + HNO_3(aq) \rightarrow NaNO_3(aq) + H_2O(l)$
D $CuO(s) + HNO_3(aq) \rightarrow Cu(NO_3)_2(aq) + H_2O(l)$
E $Mg(s) + HNO_3(aq) \rightarrow Mg(NO_3)_2(aq) + H_2(g)$

11.32 In the reaction in E above (11.31) the
A magnesium must be finely divided
B nitric acid must be very strong
C nitric acid must be very weak
D mixture must be strongly heated
E hydrogen must be removed for the reaction to proceed

11.33 Sodium hydroxide is called a deliquescent substance because it
A loses moisture to the air
B has a soapy feel
C is used as a drying agent
D absorbs moisture from the air and dissolves in it
E obtains moisture from the air but does not change its state

11.34 Ammonia gas is dried with
A calcium carbonate
B concentrated sulphuric acid
C quicklime
D anhydrous calcium chloride
E dilute sulphuric acid

11.35 A gas was tested as follows
(i) it had no reaction with limewater solution
(ii) it extinguished a lighted taper
(iii) it was colourless and odourless
The gas is most likely to be
A hydrogen
B carbon dioxide
C ammonia
D nitrogen
E carbon monoxide

11.36 Nitrogen is removed from soil as nitrates by
A bacteria
B the roots of plants
C water
D osmosis
E diffusion

Classification

Common processes used in chemistry are
A neutralisation
B oxidation
C electrolysis
D catalysis

State the process used in each of the following
11.37 the production of chlorine from brine
11.38 the preparation of sodium sulphate
11.39 the preparation of ammonia in the Haber process
11.40 the preparation of chlorine from hydrogen chloride

Multiple completion

11.41 Some hydrogen was prepared in a flask and a few test tubes of the gas collected. One tube of gas was tested with a lighted taper. An explosion was heard and a few drops of a colourless liquid was formed.

This is because
1 the flask contains air
2 the colourless liquid was water
3 hydrogen explodes when lit

11.42 When concentrated sulphuric acid is àdded to water
1 heat is evolved
2 the mixture formed is an electrolyte
3 the acid was neutralised

Assertion–reason

11.43 When burned in excess oxygen diamond and graphite produced different oxides of carbon **because** Diamond and graphite are different forms of carbon

11.44 Hard water is found in limestone areas **because** Limestone is a form of calcium carbonate

Structured questions

11.45 (a) Describe the chemical reactions involved in the industrial preparation of sulphuric acid

(b) Give the name and formula of a fertilizer that is manufactured from sulphuric acid. Give the chemical reaction involved in its formation

(c) Describe what is observed in the following experiments and state the type of chemical reaction involved in **each** case

(i) Concentrated sulphuric acid is added to sugar ($C_{12}H_{22}O_{11}$)

(ii) Potassium nitrate is heated with concentrated sulphuric acid

(iii) Dilute sulphuric acid is added to aqueous barium chloride

11.46 (a) Describe how nitrogen and oxygen are produced on an industrial scale from air

(b) How would you show that exhaled air contains carbon dioxide?

(c) Describe and explain what happens when, using a test pipette, a small volume of liquid bromine is placed in the bottom of a gas jar of air and the lid closed

(d) When magnesium is burned in air, it combines with both oxygen and nitrogen

(i) Write an equation for the reaction between magnesium and oxygen

(ii) In the reaction with nitrogen, a solid magnesium nitride, is formed, and 7.2 g of magnesium were found to give 10 g of magnesium nitride ($N=14$, $Mg=24$). Deduce the empirical (or simplest) formula for magnesium nitride

(iii) Magnesium nitride reacts with water to give ammonia gas and magnesium hydroxide. Write an equation for the reaction and state how you would verify that the gas was ammonia

Chapter 12
Organic Chemistry

Simple multiple choice

12.1 Ethane is
A an alkene
B a hydrocarbon
C an unsaturated organic compound
D unable to burn in air
E an alkyne

12.2 If a test tube of ethene is shaken up with bromine, the bromine is decolourised because ethene is
A a gas
B an inorganic compound
C an unsaturated compound
D a saturated compound
E an isomer

12.3 If 1 mole of each of the following alcohols burned in excess oxygen, the most heat would be evolved by
A ethanol B propanol C butanol D pentanol

12.4 The main constituent of natural gas is
A methane B ethane C propane D butane
E octane

12.5 In making certain sauces for steak dishes, brandy is sometimes poured into the hot frying pan containing the steak and vegetables. The brandy ignites because
A the brandy undergoes a chemical change and one of the products formed is flammable
B alcohol fumes are produced which burn readily
C water is lost from the brandy when added to the frying pan
D alcohol only burns above a certain temperature

12.6 The formula for ethyne is
A C_2H_2 B C_3H_6 C CH_4 D C_4H_{10} E C_6H_6

12.7 The pair that are isomers is
A methanol and ethanoic acid

B ethanol and dimethyl ether
C propane and butane
D dichloromethane and monochloroethane
E ethanol and propanol

12.8 The general formula for alkanes is
A C_nH_{2n} B C_nH_{2n+1} C C_nH_{2n+2}
D C_nH_{2n+3} E C_nH_{2n+4}

12.9 The hydrocarbon which is **not** an alkane is
A CH_4 B C_2H_6 C C_3H_6 D C_4H_8
E C_5H_{10}

12.10 In the formation of ethyl ethanoate the equation is

$$C_2H_5OH + CH_3COOH \rightleftharpoons CH_3COOC_2H_5 + H_2O$$

The reaction will proceed to the right if
A platinum is used as a catalyst
B the concentration of the reactants is increased
C the reactants are heated
D the water formed is removed
E the pressure is increased

Classification

A bakelite
B ethanoic acid
C ethanol
D ethene
E ethyl ethanoate
F ethyne
G hydrogen
H polystyrene

From the list name
12.11 two substances obtained directly from crude oil
12.12 an ester
12.13 a monomer
12.14 a thermosetting plastic
12.15 a thermosoftening plastic
12.16 a substance that has a treble (triple) bond

Multiple completion

12.17 The substance obtained by the fractional distillation of crude oil is
1 lubricating oil

2 petrol
3 petroleum

Assertion–reason

12.18 Methane is a member of the homologous series called the alkanes **because** Methane burns in air to form carbon dioxide and water

Chapter 13
Radioactivity

Simple multiple choice

13.1 Which of the following consists of a stream of positively charged particles?
A gamma rays
B X-rays
C alpha rays
D beta rays
E infrared rays

13.2 When the radioactive atom $^{238}_{92}Y$ decays by the emission of a β particle into the element X the product will be
A $^{234}_{90}X$ B $^{238}_{93}X$ C $^{234}_{92}X$ D $^{238}_{90}X$ E $^{234}_{94}X$

13.3 A radioactive element P decays by the emission of an α particle into element Q. It is now possible to say from this information that
A the atomic mass of Q is greater than that of P
B element Q is radioactive
C the atomic mass of P is greater than that of Q
D the rate of decay of P is greater than that of Q
E Q decays by emission of β particles

13.4 Radioactive substances give out three types of emissions, α particles, β particles and γ rays. Of these the most penetrating are
A α particles
B β particles
C γ rays
D α particles and β particles (the same)
E α particles and γ rays (the same)

13.5 Which diagram is correct?

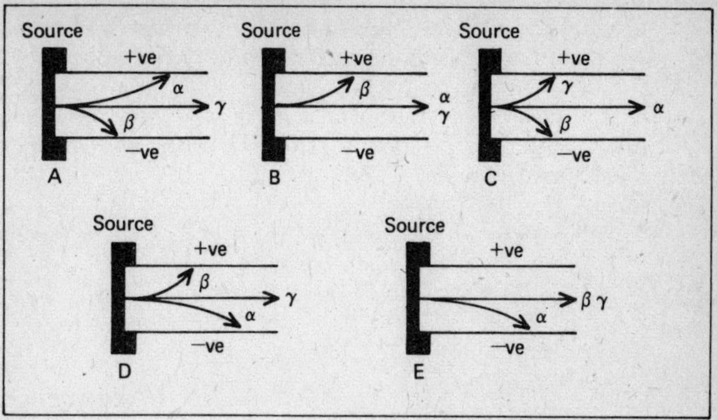

Figure 14

13.6 $^{238}_{92}W \rightarrow ^{234}_{90}X \rightarrow ^{234}_{91}Y \rightarrow ^{234}_{92}Z$
In which order are α, β and γ rays given off to convert W into Z?
A α γ β B α β β C β α β D β γ β E α β γ

13.7 A radioactive element has a half life of 5 years. If 100 g of X are left for 10 years how many g of X would remain?
A 0 g B 10 g C 20 g D 25 g E 50 g

13.8 The radioactive process is affected by
A strong heat
B use of catalyst
C pressure
D concentration
E nothing

Structured question

13.9 (a) For the named emissions which occur in radioactivity, give

	alpha particles	beta particles
(i) the masses (ii) the charges		

(b) (i) For the three stages of the radioactive decay of the isotope of radium, of mass number 228, enter in the appropriate spaces the **atomic numbers, mass numbers** and **types of radiation emitted** (α or β)

$$^{228}\text{Ra} \rightarrow {}^{228}_{89}\text{Ac} \rightarrow \ \text{Th} \rightarrow {}^{224}\text{Ra}$$

(ii) Give the composition of the nuclei of the two isotopes of radium ^{228}Ra and ^{224}Ra.

Chapter 14
Qualitative Analysis

Simple multiple choice

14.1 A lilac flame in a flame test indicates the presence of
A sodium
B calcium
C potassium
D lithium
E copper

14.2 Iron(II) and iron(III) compounds can be best identified by
A the colour of their salts
B the way in which they corrode
C adding sodium hydroxide solution
D their solubility in water
E adding dilute hydrochloric acid

14.3 All acids produce effervescence with
A sulphates
B carbonates
C sulphites
D nitrates
E phosphates

14.4 An acid is best and most easily identified by its
A sour taste
B reaction with red litmus
C reaction with blue litmus
D reaction with metals
E reaction with an alkali

14.5 Sulphur dioxide can best be identified by its
A appearance
B solubility in water
C reaction with potassium dichromate(VI)
D reaction with litmus
E smell

14.6 The acid gas which produces a white smoke with ammonia is

A sulphur dioxide
B hydrogen chloride
C sulphur(VI) oxide (sulphur trioxide)
D chlorine
E nitrogen dioxide

14.7 When tested with a lighted taper hydrogen will explode if it
A is hot
B mixed with air
C is cooled
D is moist
E is dry

Classification

A chloride
B bromide
C iodide
D sulphate
E carbonate

The test result corresponding to each radical is
14.8 a white precipitate soluble in aqueous ammonia
14.9 a white precipitate that effervesces when an acid is added to it
14.10 a pale yellow precipitate sparingly soluble in aqueous ammonia
14.11 a yellow precipitate insoluble in aqueous ammonia

Multiple completion

14.12 When tested with moist litmus paper chlorine
1 bleaches the litmus paper
2 turns the litmus paper red
3 turns the litmus paper blue

14.13 Fluorine is
1 more reactive than chlorine
2 has a higher atomic number than iodine
3 a colourless solid

Chapter 15

Plastics and Fibres

Simple multiple choice

15.1 Polymers are made from
A alkanes
B isomers
C allotropes
D monomers
E particles

15.2 A thermoplastic is one that
A hardens with heat
B eventually melts with very strong heat
C softens when heated
D is made using heat
E disintegrates when heated

15.3 Ropes used by climbers are often made of
A nylon
B terylene
C perspex
D PTFE
E polyvinyl chloride (PVC)

15.4 One major advantage of plastics over wood, metal and stone is that they
A require regular painting
B do not deteriorate
C can be used for every purpose
D are denser than the other materials
E are always easy to shape

15.5 Nylon is manufactured using a condensation polymerisation process. This means that
A water is eliminated in the process
B water is added during the process
C water combines with the reactants
D other liquids are formed in the process
E steam is formed and condenses in the reaction

Multiple completion

Below is a list of well known plastics

A Teflon (PTFE)
B Bakelite
C Polyurethane
D Melamine

Choose the plastic most suitable for each of the following uses

15.6 paints
15.7 tableware
15.8 electrical components
15.9 non stick surfaces
15.10 brake linings

Answers

Chapter 1

1.1C, 1.2B, 1.3A, 1.4B, 1.5B, 1.6D, 1.7B, 1.8A, 1.9D, 1.10C, 1.11E, 1.12C, 1.13E, 1.14D, 1.15E, 1.16C, 1.17A, 1.18B, 1.19A, 1.20C, 1.21D, 1.22B, 1.23A, 1.24E, 1.25B, 1.26A.

1.27 (a) (i) The limewater would turn milky in each case, most milkiness appearing in the bottom jar because most of the carbon dioxide would be present there because it is more dense than air. By the process of diffusion some carbon dioxide would be in the top gas jar
(ii) 'Pops' would be produced in both gas jars because of the air and hydrogen diffusing between each other.
(b) Bromine is a very volatile liquid
(c) The process of diffusion
(d) Carbon dioxide does not diffuse as rapidly as hydrogen.

Chapter 2

2.1C, 2.2D, 2.3B, 2.4C, 2.5A, 2.6E, 2.7C, 2.8D, 2.9D, 2.10A, 2.11E, 2.12B, 2.13C, 2.14C, 2.15E, 2.16D, 2.17A, 2.18B, 2.19C, 2.20 1. A&C, 2. C&C, 3. C&B, 4. A&A, 5. B&C, 2.21D, 2.22A, 2.23C, 2.24B, 2.25F, 2.26D, 2.27B, 2.28B, 2.29D, 2.30C, 2.31A, 2.32A.

Chapter 3

3.1C, 3.2A, 3.3D, 3.4D, 3.5A, 3.6D, 3.7D, 3.8B, 3.9A, 3.10C, 3.11B, 3.12D, 3.13C, 3.14E, 3.15A, 3.16D, 3.17B, 3.18B, 3.19B, 3.20D.

Chapter 4

4.1B, 4.2A, 4.3B, 4.4D, 4.5C, 4.6D, 4.7D, 4.8B, 4.9B, 4.10A, 4.11E, 4.12B, 4.13C, 4.14B, 4.15C, 4.16E, 4.17B, 4.18C, 4.19D, 4.20C, 4.21A, 4.22A, 4.23B, 4.24C, 4.25D, 4.26C, 4.27B, 4.28B, 4.29A, 4.30A, 4.31B, 4.32B, 4.33B, 4.34E, 4.35B, 4.36B.

4.37 (a) 1000 cm^3 of M silver nitrate solution contains 1 mole
100 cm^3 of M silver nitrate solution contains 1/10 mole mole
100 cm^3 of 0.5 M silver nitrate solution contains 1/20 mole
(b) From the equation
2 moles of silver nitrate react with 1 mole of the bromide

1/20 mole of silver nitrate react with 1/40 mole of bromide

(c) 1/40 mole of $MBr_2 = 4.6$

1 mole of $MBr_2 = 4.6 \times 40 = 184$

(d) $MBr_2 = 184$, $Br_2 = 160$ (i.e. 2×80), $M = 184 - 160 = 24$

Chapter 5

5.1B, 5.2C, 5.3C, 5.4B, 5.5A, 5.6D, 5.7B, 5.8A, 5.9C, 5.10C, 5.11D, 5.12C, 5.13B, 5.14E, 5.15B, 5.16C, 5.17B, 5.18B&D, 5.19D, 5.20E, 5.21A, 5.22A&C, 5.23B, 5.24 A-lead bromide, B-copper, C-silicon dioxide, D-graphite, E-magnesium oxide, 5.25B, 5.26C, 5.27B, 5.28C, 5.29A.

5.30 (a) (i) 2, 8, 2, (ii) 2, 8, 7.

(b) Sodium has the electron arrangement 2, 8, 1, while chlorine has the electron arrangement 2, 8, 7. Sodium donates an electron to chlorine. This means that sodium now has 10 electrons + 11 protons, while chlorine now has 18 electrons and 17 protons. Both atoms are now charged, Na^+, Cl^- and are now ions.

(c) The bonding in solid sodium chloride is the result of the attraction between the ionic pair Na^+ and Cl^-. In a solid the packing is very close, consequently the closer the pair, the greater the attractive forces. Melting the compound breaks down the solid state into the liquid state, the ionic pair are further apart, attraction is weaker, consequently decomposition via electrolysis is easier and achievable.

(d) (i) 2, 8, 8, (ii) Neon, Ne 2, 8 (or any other noble gas) (iii) They all have complete outer shells of electrons.

(e) (i) 2, 8, 5 (ii) PH_3, covalent bonding (see ammonia)

(f) (i) MgO, Al_2O_3, (ii) basic, amphoteric.

Chapter 6

6.1A, 6.2C, 6.3B, 6.4B, 6.5D, 6.6B, 6.7B, 6.8D, 6.9D, 6.10B, 6.11B, 6.12A.

6.13 (a) (i) Oxidation is the gain of oxygen or the loss of hydrogen. Reduction is the reverse of this, i.e. the loss of oxygen and the gain of hydrogen (ii) Oxidation is the loss of electrons by a substance. Reduction is the gain of electrons by a substance.

Note: In all examinations it is good practice to give examples. The lack of space in this answer section precludes full answers and examples in many instances.

(b) (i) $Zn(s) + Pb(NO_3)_2(aq) \longrightarrow Pb(s) + Zn(NO_3)_2(aq)$.

The products are lead and zinc nitrate solution. The zinc is added to the lead nitrate solution under normal laboratory conditions of temperature and pressure. The zinc has been oxidised to zinc nitrate, while the lead nitrate has been reduced to lead.

i.e. zinc has lost electrons $Zn - 2e^- \rightarrow Zn^{2+}$, lead has gained electrons $Pb^{2+} + 2e^- \rightarrow Pb$.

(ii) $CuO(s) + CO(g) \rightarrow Cu(s) + CO_2(g)$

Copper and carbon dioxide are the products. The carbon monoxide is passed over heated copper oxide. Care must be taken to ensure that the carbon monoxide is dispersed safely (**poisonous!**). The copper oxide is reduced by the carbon monoxide to copper, i.e. it has lost oxygen. Carbon monoxide is oxidised to carbon dioxide, i.e. it has gained oxygen.

(c) Both iron and aluminium corrode, both forming oxides. The aluminium oxide film is tenacious, clinging to the metal thereby preventing further aluminium being exposed to air. Not so iron oxide which when formed flakes off the metal exposing fresh metal for the corrosive attack.

6.14 (a) Answered in Q.6.13. (b) Reaction A represents an oxidation/reduction action because the magnesium loses electrons

$Mg - 2e^- \rightarrow Mg^{2+}$ + reduction and the hydrogen ions gain electrons $2H^+ + 2e^- \rightarrow H_2$ + oxidation.

(c) $PbBr_2 \rightleftharpoons Pb^{2+} + 2Br^-$

At the anode (+ve) $2Br^- - 2e^- \rightarrow Br_2$ – oxidation

At the cathode (–ve) $Pb^{2+} + 2e^- \rightarrow Pb$ – reduction

Chapter 7

7.1B, 7.2B, 7.3D, 7.4B, 7.5C, 7.6B, 7.7E, 7.8C, 7.9B, 7.10D, 7.11E, 7.12D, 7.13C, 7.14A, 7.15C, 7.16B, 7.17B, 7.18D, 7.19D, 7.20D, 7.21D, 7.22A, 7.23E, 7.24C, 7.25D, 7.26B, 7.27A, 7.28B, 7.29B, 7.30A, 7.31A,

7.32 (a) D (b), B (c) Sodium or potassium because they react with water (making the solutions) (d) D, E, A, C, B (e) (i) Copper, lead, iron, magnesium (ii) oxygen because all are aqueous solutions, resulting in OH^- ions being discharged at the anode in every case.

7.33 (a) A (b) The formula for lead(II) bromide is $PbBr_2$. The ions present are Pb^{2+} and $2Br^-$.

(c) The negatively charged Br^- ions will travel to the positively charged anode B because unlike charges attract each other.

(d) Anode: $2Br^- \rightarrow Br_2 + 2e^-$

Cathode: $Pb^{2+} + 2e^- \rightarrow Pb$

(e) At the anode B a reddish brown gas, bromine, is seen which condenses upon cooling, into a dark red liquid. At the cathode there is little to see except an intense glow around the electrode. Upon cooling the cathode is seen to be coated with a grey metal – lead.

Chapter 8

8.1D, 8.2D, 8.3A, 8.4C, 8.5C, 8.6C, 8.7C, 8.8C, 8.9(i)A, (ii)E, (iii)D, (iv)A, (v)C or B, 8.10A, 8.11C, 8.12D, 8.13E, 8.14D, 8.15E, 8.16D, 8.17A, 8.18E, 8.19A, 8.20C, 8.21B & E, 8.22C, 8.23C, 8.24A or B, 8.25B.

8.26 (a) Litmus/methyl orange,
(b) $2NaOH(aq) + H_2SO_4(aq) \rightarrow Na_2SO_4(aq) + H_2O(l)$ (c) 1/20, (d) 12.5 cm^3, (e) The experiment should be repeated **without** the indicator otherwise the crystals obtained will be coloured.

8.27 (a) Effervescence/bubbling, (b) To remove any undissolved zinc carbonate (c) 1/20 mole = 1/20 of the molar mass of zinc carbonate $1/20 \times 165 = 8.25$ g. (d) (i) Allow the precipitate formed initially to settle, then add a few drops of lead nitrate to the supernatant liquid on the top of the precipitate (ii) Filtration or centrifuging (iii) To free it from the zinc nitrate solution (iv) sodium or potassium sulphate – any soluble sulphate.

Chapter 9

9.1B, 9.2C, 9.3C, 9.4D, 9.5D, 9.6D, 9.7B, 9.8D, 9.9B, 9.10B, 9.11C, 9.12D, 9.13A, 9.14E, 9.15C, 9.16B, 9.17A, 9.18B, 9.19B, 9.20B, 9.21B, 9.22B.

9.23 (a) Sodium, Na (b) Neon, Ne, (c) Sulphur, (d) Its compounds would be coloured, and it would form two series of compounds Mn(II) and Mn(IV), (e) (i) LiCl, (ii) white, (iii) soluble, (iv) conductor, (f) AsH_3.

Chapter 10

10.1C, 10.2A, 10.3E, 10.4A, 10.5D, 10.6C, 10.7B, 10.8C, 10.9C, 10.10B, 10.11E, 10.12D, 10.13B, 10.14B, 10.15C, 10.16C, 10.17C, 10.18A, 10.19D, 10.20A, 10.21C, 10.22D, 10.23B, 10.24A, 10.25A, 10.26D, 10.27D, 10.28A, 10.29B, 10.30B, 10.31B, 10.32A, 10.33D, 10.34A, 10.35D, 10.36B, 10.37C, 10.38B, 10.39A, 10.40C,

10.41 (i) Calcium carbonate – white, (ii) iodine – grey, (iii) copper(II) sulphate – blue, (iv) iron(II) sulphate – green, (v) silver iodide – black, 10.42 (i) aluminium, (ii) lead, (iii) zinc, (iv) copper, (v) carbon, 10.43 (i) sodium, (ii) calcium, (iii) copper, (iv) zinc, (v) lead, 10.44B, 10.45B, 10.46A, 10.47(a)D, (b) A: there has according to the table been an increase in weight after heating;

in this instance there should have been a decrease, (c) The reaction is complete, (d) 1.60 g, (e) $CuCO_3(s) \rightarrow CuO(s) + CO_2(g)$, 1 mole of copper carbonate produces 1 mole of copper oxide

i.e. 124 g of copper carbonate produce 80 g of copper oxide
248 g of copper carbonate produce 160 g of copper oxide
2.48 g of copper carbonate produce 1.60 g of copper oxide

10.48 (a) (i) hydrogen, (ii) Na^+, Cl^-, H^+, OH^-, (iii) it does not corrode, (iv) on the coast, easy and economical access to the raw material.

(b) (i) copper, (ii) aluminium, 10.49 (a) water, air(oxygen), (b) C, (c) G, (d) C, (e) D, (f) B, (g) oxidation, (h) 160, 10.50(a) (i) sodium nitrate – no reaction, zinc nitrate – no reaction, copper(II) nitrate – brown copper is precipitated and the blue colour gradually becomes lighter.

(ii) Zinc carbonate – changes from white to yellow, and back to white again when the test tube cools down. Sodium carbonate – no reaction. Copper(II) carbonate, changes colour from green to black, (iii) Zinc – effervescence occurs with the evolution of hydrogen gas. Copper – no reaction occurs. Sodium, zinc, lead, and copper is the order of decreasing activity because (i) neither sodium nor zinc were displaced by lead or copper. Sodium carbonate was stable when heated while zinc carbonate decomposed. Lead displaced copper from copper nitrate,

(b) (i) $FeSO_4 7H_2O(S) \rightarrow FeSO_4(s) + 7H_2O(l)$

$$2FeSO_4(s) \rightarrow Fe_2O_3(s) + SO_2(g) + SO_3(g)$$

The water of crystallisation is first evolved and the compound changes from pale green to white. Clouds of acid gases are evolved (SO_2 and SO_3) when heating is continued. A brown residue of iron(III) oxide remains in the tube, (ii) The water of crystallisation is given off and the blue crystals change to a white powder. With normal heating no further reaction takes place

$$CuSO_4.5H_2O(s) \rightarrow CuSO_4(s) + 5H_2O(l)$$

Blue crystals   white powder

(iii) Strong heat melts the sodium nitrate, oxygen gas is given off, and sodium nitrite is the other product

$$2NaNO_3(s) \rightarrow 2NaNO_2(s) + O_2(g)$$

Chapter 11

11.1A, 11.2C, 11.3C, 11.4D, 11.5B, 11.6C, 11.7C, 11.8D, 11.9E, 11.10B, 11.11C, 11.12C, 11.13C, 11.14A, 11.15A, 11.16C, 11.17D,

11.18B, 11.19B, 11.20D, 11.21C, 11.22C, 11.23C, 11.24D, 11.25B, 11.26C, 11.27D, 11.28C, 11.29B, 11.30C, 11.31A, 11.32C, 11.33D, 11.34C, 11.35D, 11.36A, 11.37C, 11.38A, 11.39D, 11.40B, 11.41B, 11.42B, 11.43A, 11.44B, 11.45(a) Sulphur dioxide produced by burning sulphur or roasting sulphide ores in air is mixed with oxygen obtained by the fractional distillation of liquid air. The gases are dried

$$S(s) + O_2(g) \rightarrow SO_2(g)$$

The gas mixture is now passed over a heated catalyst of vanadium(V) oxide (vanadium pentoxide), and sulphur(VI) oxide (sulphur trioxide) is produced

$$2SO_2(g) + O_2(g) \rightarrow 2SO_3(g)$$

The sulphur trioxide is bubbled into strong sulphuric acid producing oleum (fuming sulphuric acid)

$$SO_3(g) + H_2SO_4(aq) \rightarrow H_2S_2O_7(l)$$

Sulphuric acid of the required strength can now be produced by diluting the oleum with water, (b) Ammonium sulphate, $(NH_4)_2SO_4$

$$2NH_3(g) + H_2SO_4(aq) \rightarrow (NH_4)_2SO_4(aq)$$

(c) (i) The sugar changes from white to yellow to dark brown to black due to the dehydrating action of concentrated sulphuric acid

$$\underset{\text{(white)}}{C_{12}H_{22}O_{11}} \xrightarrow{\text{dehydration}} \underset{\text{(black)}}{12C + 11H_2O}$$

(ii) Fumes of nitric acid are produced which condense on the cool parts of the tube. In this instance the less volatile sulphuric acid displaces the more volatile nitric acid from its salt.

$$KNO_3(s) + H_2SO_4(l) \rightarrow KHSO_4(aq) + HNO_3(g)$$

(iii) A white precipitate of barium sulphate is formed. This is an example of precipitation, and is also a test for a sulphate.

$$H_2SO_4(aq) + BaCl_2(aq) \rightarrow BaSO_4(s) + 2HCl(aq)$$

11.46 (a) Very briefly, air is subjected to pressure and cooled until it becomes liquid. This liquid air is a mixture of the various gases found in air, and they can be separated by virtue of their different boiling points. Fractional distillation produces each gas in turn.